NOISE AND THE SOLID STATE

NOISE AND THE SOLID STATE

D. A. Bell
Professor Emeritus of Electronic Engineering,
University of Hull

PENTECH PRESS
London : Plymouth

First published 1985
by Pentech Press Limited,
Estover Road, Plymouth,
Devon PL6 7PZ

British Library Cataloguing in Publication Data

Bell, D. A.
  Noise and the solid state.
  1. Electronic noise
  I. Title
  621.38'0436   TK7867.5

  ISBN 0-7273-1405-X

Printed and bound in Great Britain

Preface

Even in the days of thermionics, 'noise' in electrical conduction was divided between two main fields: variation of velocity of charge carriers associated with thermal noise and variations in number associated with shot noise. Views on thermal noise have been little affected by the change to solid-state devices, though it is now coming to be known as diffusion noise, the links with temperature and conductivity being maintained by the Einstein relationship between diffusion, mobility and temperature. On the variable-number side, shot noise in the thermionic vacuum tube is replaced by injection noise in the solid-state device, together with generation-recombination noise and the effects of traps. The $1/f$ noise which was first observed in semiconductors (but has now been found also in metals) is still the subject of controversy. In addition semiconductors have brought the new phenomena of hot electrons, avalanche multiplication and transferred-electron (Gunn effect) devices.

Apart from the incompletely solved problem of $1/f$ noise, the growing points of the subject appear to be in the use of cryogenic devices and in radiation detectors. These two are linked in the attempt to bridge the gap between thermal devices for the far infrared and radio type detectors for sub-millimetre wavelengths, and in the attempt to achieve sufficient sensitivity to detect gravitational radiation.

It is hoped that it is now possible to provide an adequate grounding in fundamental principles for whatever direction the development of theory and of technological applications may take. Some existing devices are necessarily described in order to illustrate the ways in which noise may arise in a system, the ways in which it may be minimised and the present frontiers of technology in this respect. The book does not set out to record the characteristics of all the solid-state devices which are currently in use, nor to provide detailed design information for them; but it gives sufficient references to the original literature to enable the reader who has special interests to follow up the development of any particular device.

D. A. Bell

Contents

Chapter 1

Thermal Noise

1.1 Introduction

Noise—the short-hand term for random fluctuations of power in an electrical system—is the limiting factor in the signalling power required in practically all communication channels (broadcasting may be an exception), at any rate if one includes noise-like interference such as 'static' and 'man-made static' in radio communication and random intermittencies in line communication. There are arguments for treating interference separately from the more fundamental forms of noise. First, if the interference is another signal or otherwise man-made, the proper remedy is to eliminate it at source. Second, if its characteristics are known, i.e. it is not random, special measures may be employed to distinguish between signal and interference which would not be effective against random noise. An example is the use of a limiter as 'spotter' to render ignition interference less noticeable in television. But if so many independent sources of interference are operating simultaneously that the interference appears random it can be handled theoretically by the same techniques which are used for physically generated random noise: for example one may attribute an increased effective temperature to a radio aerial in order to take account of various kinds of interference which it picks up. (See Section 1.9 for the temperature of the radiation resistance of an aerial.) Communication theorists then use the diagram of Fig. 1.1 to represent the position *if the receiver adds no noise*. The original signal S' and the noise N' are changed in level to some predictable extent by the channel, so as to arrive at the receiver with values S and N. This diagram can be justified by noting that from the point of view of communication theory the 'detector' is the critical part of the receiver (because it is usually non-linear and operates in such a way as to decrease the amount of available information and change the signal-to-noise ratio) and the pre-detector amplifiers can be treated as part of the channel. This book, however, is concerned precisely with the noise generated in those amplifiers, so that a diagram of the form of Fig. 1.2 is more appropriate. The tendency in

1

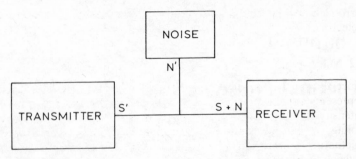

Fig. 1.1 Conventional equivalent system of communication theory

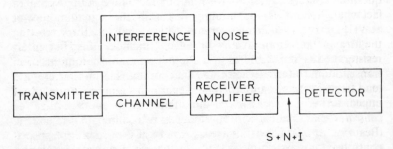

Fig. 1.2 Diagram of communication system, showing noise generated in receiver

recent years has been to use higher frequencies, with a reduction in interference but an increase in the noise arising within the receiver. There is now a discontinuity between the use on the one hand of radio signals of 1 to 10 GHz (or sometimes up to 30 GHz) for satellite communications, or 30 to 100 GHz in low-loss waveguides and, on the other hand, of optical signals for transmission through glass fibres. One thinks of entirely different techniques for the reception of the latter type of signal—and, indeed, the impact of quantum theory on optical signals is such as to justify a separate treatment of 'radiation detectors' in Chapter 8. In every case, however, the requirement is to produce a detectable electric current in response to an incoming signal—usually an electromagnetic signal, either radio or optical, but possibly gravitational.

Chapter 2 is devoted to $1/f$ noise, an additional type of noise found in semiconductors of which the mechanism is still controversial. Chapters 3 to 8 deal with noise in particular types of device or system, some of which involve types of noise other than those found in electrical conductors, e.g. Barkhausen noise in ferromagnetic and ferro-electric materials and photon fluctuations in radiation detectors.

1.2 Noise power

Since the fluctuation of electric current, which is viewed as noise, represents a stochastic phenomenon, it must be described in terms of intensity through a mean-square parameter, e.g. a mean-square voltage or current, and the spectral distribution of the squared fluctuation is known as a 'power spectrum'. The latter may also be *defined* as the Wiener–Khinchine transform of the autocorrelation function of the time representation of the stochastic phenomenon in question. It is the signal-to-noise *power* ratio which is significant, a fact which became obvious when thermionic devices came to be used at VHF and again when solid-state devices were first introduced in the form of the junction transistor which has a relatively low input resistance. It might be thought that the transition to field-effect transistors, through the several variants of MOS technology, would tend to obscure the importance of power because of their high input impedance. The opposite can be shown through the concept of *charge control* which was first introduced in relation to junction transistors (Beaufoy and Sparkes, 1957) but is equally applicable to grid-controlled thermionic valves or FETs. It is not immediately apparent that there is any dissipation of energy, and hence of power, necessarily associated with charge, as there is with current or voltage in a circuit of finite resistance. For, if one thinks of a perfect FET having infinite input resistance, and working at a frequency low enough to make dielectric loss insignificant, the input impedance is purely capacitive and therefore unable to absorb power. But the presence of charge on this capacitance involves a stored energy of electrostatic field of amount $\frac{1}{2}\varepsilon E^2$ per unit volume of the dielectric, or in circuit form $\frac{1}{2}QV$, where Q is the charge and V the potential at which Q is stored. To build up this stored energy in a given time requires a certain power input and to remove it requires dissipation; and the handling of any kind of information-carrying signal requires that it should be possible to change the stored energy at an arbitrary rate. (Energy can be taken in and out of a capacitance without loss in a resonant circuit, but only at a single frequency which does not constitute an information-carrying signal.)

1.3 Equipartition

The concept of power as the rate of change of energy is further important because one ubiquitous source of noise, thermal noise, is related to energy through the equipartition theorem which states that every physical system at temperature T contains energy of average amount $\frac{1}{2}kT$ per degree of freedom, where k is Boltzmann's constant. 'Degree of freedom' is a difficult concept in the abstract, but generally the number of degrees of freedom is equal to the number of variables needed to specify the state of the system. A spherical particle moving freely in three dimensions has three degrees of freedom and a harmonic oscillator has two. The conditions for the validity of the equipartition theorem can be derived from statistical mechanics (Tolman, 1938) and can be summarised as follows:

(1) Use of the 'frequency' theory of probability. (This comes in through counting of what are sometimes called the 'complexions' of a system.)
(2) Conservation of number and of total energy.
(3) Energy associated with a degree of freedom proportional to the square of the relevant variable, e.g. kinetic energy of a particle proportional to the square of velocity, energy stored in a capacitance proportional to square of voltage, energy stored in an inductance proportional to square of current.

The microscopic physical mechanism of equipartition in electrical conductors is described in Section 1.4. The transformation from *total average energy* (including components of fluctuation at all frequencies), which is the factor governed by equipartition and customarily considered in thermodynamics, to *power in a given frequency range*, which is the factor of concern to communication engineers, was achieved by Nyquist in 1928 and will also be discussed in Section 1.5 when noise is examined from a circuit viewpoint.

1.4 The microscopic mechanism of thermal noise

Since the mean value of electric current (other than displacement current) can be represented by

$$\bar{J} = \overline{qnv} \tag{1.1}$$

where J is current density, q the charge per mobile particle, n the numerical density of particles and v their velocity, fluctuations in current can arise from either fluctuations in n or fluctuations in v (assuming that q is constant, as for electrons and holes).

But the effective mass of an electron (or hole) in a particular energy state in a semiconductor may be very different from the mass of a free electron. In Gunn or transferred-electron devices some conduction electrons may transfer from one energy state or 'valley' to another, with change of velocity for a given applied electric field following the change in effective mass. It has been shown (Burgess, 1959) that if n and v are uncorrelated

$$\text{var}(J) = \bar{n}\,\text{var}(v) + \bar{v}\,\text{var}(n) \tag{1.2}$$

where var(J) stands for variance of J, etc. (The variance is the mean square of the departure from the mean; and in electrical applications it is the mean power in the a.c. component after the mean or d.c. component has been eliminated, e.g. by a blocking capacitor. It is often assumed that the mean is thus zero, so that the variance can be replaced by the mean square.) The right-hand side of (1.2) can be interpreted as the sum of thermal noise and number-fluctuation noise, the simplest form of the latter being shot noise which is examined in Appendix 1. There is also generation-recombination noise in a semiconductor which is a variable-number effect, though with a Lorentzian rather than white (shot noise) spectrum; and the question whether $1/f$ noise is due to fluctuation in number of charge carriers is still in dispute. The variance of J still exists when the mean $\bar{J} = 0$, but in any kind of conductor at zero mean current $\bar{v}$ is also zero, so the second term in (1.2) then vanishes, leaving only the thermal noise which is coming to be known as diffusion noise*. The latter term is logical, since it is the random movements which are responsible for diffusion and the quantitative relationship between diffusion and temperature is expressed in the Einstein formula

$$qD = \mu kT \tag{1.3}$$

where D is the diffusion constant, μ the mobility, k Boltzmann's constant and T the temperature. (For an early discussion of diffusion noise see Van der Ziel and Van Vliet, 1968.) If there is no steady current, or if the conductor is linear and homogeneous so that the mean current can be subtracted from the total, the noise current density can be written

$$\Delta J^2 = \bar{n}\,\Delta v^2 + \bar{v}\,\Delta n^2 \tag{1.4}$$

It is generally assumed that in metals n is constant, there being approximately one conduction electron per atom. The number of charge carriers and their mobility in a material in which they are predominantly of one sign can be determined by measuring both the conductivity, which is proportional to the product of mobility and number, and the Hall coefficient, which is proportional to their ratio. (The mobility so determined is called the Hall mobility, as distinct from the mobility determined by observing the transit time of a pulse

---

* There is also $1/f$ noise (Chapter 2) which is usually observed in the presence of a mean current, though there is evidence that it still exists in the absence of mean current, in which case it cannot be due to number variation.

Table 1.1

Metal	Valency	No. of electrons in outermost shell	Mobile electrons per atom
Gold	1 or 3	1	1.5
Silver	1	1	1.3
Copper	1 or 2	1	1.1

of current along a semiconductor. This transit-time method is only applicable to semiconductors because the pulse must be of injected carriers—in a metallic conductor one observes the propagation velocity of the electric field, not of the carriers.) It is sometimes assumed that the mobile electrons in a metal are identical with the valence electrons, but the numbers of mobile electrons per atom found experimentally by Ehrenberg (1958)* vary somewhat with temperature and are not integers. The values for gold, silver and copper at room temperature (17°C or 290 K) are given in Table 1.1. Appreciably smaller values were found at 15 K. The variation of number with temperature and the departure from integral values will have to be considered again in connection with $1/f$ noise; but, as thermal noise involves only the mean value of n, it is not affected by the question whether n is strictly constant or not.

The mention of diffusion brings to mind Langevin's transport equation, which in terms of the density of electric current due to motion of electrons may be written

$$J_n = q\mu_n E + qD_n \, \partial n/\partial x + H(x, t) \qquad (1.5)$$

(The subscript n indicates that the parameter is to have the value relevant to electrons.) The first term represents the drift under the applied field E while the second term represents the mean effect of diffusion down the concentration gradient, a gradient which may exist in semiconductor devices of inhomogeneous structure, e.g. junction diodes. The third term represents any other force which may be operative in the system and may include a noise source. Van der Ziel and Van Vliet (1968) identified $H(x, t)$ with the fluctuations inherent in the diffusion process and used the equation to justify the separation of d.c. (drift) and thermal noise: this separation is inherent in the additive combination of terms on the right-hand side of (1.5).

There is a very general fluctuation-dissipation theorem (Kubo, 1966) which states that every source of fluctuation (of power) is

---

* Ehrenberg gives numerical densities of electrons. The corresponding numerical densities of atoms have been calculated from atomic weights and specific gravities.

associated with a mechanism of dissipation. (This does not apply directly to fluctuations of stored energy.) An example which is often used is the mechanism of Brownian motion of particles suspended in a fluid. A particle is set in motion by the transfer of momentum from the fluid molecules which are moving in accordance with their thermal kinetic energy, but the motion of the particle is damped by the viscosity of the fluid, i.e. the *average* transfer of momentum resulting from many molecular collisions. The principle is familiar in electric circuits in the form that (thermal) *noise* is associated with *resistance* in the Nyquist formulation; and it is interesting to see that this relationship can be established from the microscopic mechanism of electrical conduction. The first point to note is that the kinetic energy of the conduction electrons in a metal has a Fermi–Dirac distribution instead of the Maxwell–Boltzmann distribution which would be appropriate to the molecules of an ideal gas. This shows up experimentally in the absence of contribution from the conduction electrons to the specific heat of a metal and theoretically depends on Pauli's exclusion principle (that not more than one particle can occupy each quantum state, or two electrons per energy state on account of the spin quantum number) and so on the ratio of number of particles to number of quantum states in a given volume. In most semiconductors, on the other hand, the density of electrons in the conduction band is so small that they have a Maxwell–Boltzmann distribution. For detailed derivation of the way in which the thermal energy of conduction electrons enters into both conductivity and fluctuations, see Lorentz (1916) and Bakker and Heller (1939), or the summary in Bell (1960). The results are as follows: The *conductivity* is given by the formula

$$\sigma = -\frac{4\pi}{3}\frac{q^2 l}{m}\int_{\mathbf{r}=0}^{\infty}\mathbf{r}^2\frac{\partial f}{\partial \mathbf{r}}\,d\mathbf{r} \tag{1.6}$$

where l is the mean free path, $\mathbf{r}$ the vector velocity and f the distribution function for number of electrons with a given velocity. Both Maxwell–Boltzmann and Fermi–Dirac statistics lead to

$$\partial f/\partial \mathbf{r} = -(m\mathbf{r}/kT)(n-\bar{n})^2$$

and Bakker and Heller (1939) showed that this would be true of any kind of statistics provided only that it depends on the frequency theory of probability and that it takes the probability of a state of energy U to be proportional to $\exp(-U/kT)$. These are the same conditions which apply to the equipartition theory of mean energy equal to $\frac{1}{2}kT$ per degree of freedom. Since conductivity is meaningful only if there is a finite average or drift velocity, microscopic analysis usually assumes that the drift velocity is only a small perturbation of the total velocity distribution and can therefore be treated by superposition. The experimental justification of this assumption is that if it were not true the conductor would not be ohmic in the sense of a linear relation between current and voltage (or current-density and electric field). Modification will therefore be needed in the case of 'hot electrons', i.e. those for which scattering of the drift velocity appreciably enhances the random velocities (Chapter 4).

To find the magnitude of *fluctuations* by the same kind of analysis, one supposes a plane set up across the conductor and counts the number of particles which would cross it in either direction. The net transfer of charge per unit area is found from the difference between the numbers n^+ travelling in the positive direction and n^- in the negative direction through an element of area dA:

$$dQ/dA = q(n^+ - n^-)/dA$$

$$= q \, dt \int_{u=-\infty}^{\infty} \int_{v=-\infty}^{\infty} \int_{w=-\infty}^{\infty} uf(u, v, w) \, du \, dv \, dw \qquad (1.7)$$

where u is the component of velocity normal to the plane, v, w the other two Cartesian components of velocity, and $f(u, v, w)$ is the distribution function. If the mean field, and hence the mean value of u, is zero, the mean current is zero but there is still a fluctuation in current density

$$\overline{\Delta J^2} = q^2 \int u^2 (\Delta f) \, d\lambda \qquad (1.8)$$

where Δf is the fluctuation in number of particles in state $d\lambda = du \, dv \, dw$. If only one component, u, is considered it can be shown (Tolman, 1938) that

$$\partial n/\partial u = -(mu/kT)\overline{(n-\bar{n})^2} \qquad (1.9)$$

This differs from (1.6) in which r is the *total* velocity, but after allowing for this and for the conversion from conductivity and current density to conductance and current, an expression is obtained which is comparable with (1.6) but includes the kT factor. Formula (1.8) refers to fluctuations of all frequencies so one now has to choose between two courses in order to compare it with the Nyquist theorem for noise in circuits which will be derived below. Bakker and Heller (1939) applied the autocorrelation function in order to obtain the spectral density of fluctuation and duly found

$$S_I(f) = 4GkT \qquad (1.9)$$

in agreement with Nyquist. The other approach (Brillouin, 1934) is to relate the electromagnetic part of the kinetic energy to square of current. The total kinetic energy of a system of many electrons is

$$U = \frac{1}{2}\sum_i mv_i^2 + \frac{\mu_0 q^2}{4\pi}\frac{1}{2}\sum_i\sum_k \frac{\mathbf{v}_i\mathbf{v}_k}{r_{ik}} \tag{1.10}$$

The first term represents the 'mechanical' kinetic energy which depends on the mass of the electron and is found to be negligible compared with the second term which represents the electromagnetic effects. If the electrons have a common drift component of velocity, i.e. there is a mean current i, the corresponding part of the second term is equivalent to an inductive effect, $U_0 = \frac{1}{2}Li^2$. Using a linear perturbation approximation the fluctuation term can be written

$$U_{\text{th}} = \frac{\mu_0 q^2}{4\pi}\frac{1}{2}\sum_i\sum_k \frac{(\mathbf{v}_i - \bar{\mathbf{y}})(\mathbf{v}_k - \bar{\mathbf{y}})}{r_{ik}} \tag{1.11}$$

From this it is found that $\overline{\Delta i_{\text{tot}}^2} = kT/L$ which is the value of all-frequency fluctuation which will be derived in circuit terms from the Nyquist value of noise in $\text{Re}(Z)$ at a given frequency when this is integrated over all frequencies.

The above is now a matter of history, but it is recorded in order to show that there is a very solid foundation in atomic physics for the Nyquist formula for noise in a resistance which was first derived on a more general thermodynamic basis. It is worth commenting on the point that the change from classical to quantum theory of electrical conduction in metals did not affect the relationship between fluctuations and resistance, a result which must hold in order to satisfy the thermodynamically established fluctuation-dissipation theorem. In a Fermi–Dirac distribution for particles in a degenerate system, all states well below the Fermi level are filled. Electrons in these states cannot take part in conduction because acquisition of drift velocity would raise them to a higher quantum state which they cannot enter because it is already full. Likewise, movement to a different position is barred, because that would be to another already-filled state*, so they cannot contribute to the random movements in space which constitute thermal noise. So the $\partial f/\partial r$ factor in (1.6), which is the same for Fermi–Dirac as for Maxwell–Boltzmann distributions, enters into both conductivity and fluctuation calculations.

1.5 Macroscopic equipartition

Another approach to the prediction of fluctuations in electric circuits—the one which was in effect founded by Nyquist in order to

---

* A 'cell' in the array of quantum states is defined by two conjugate co-ordinates, i.e. two coordinates of which the product has the dimensions of action, and two such co-ordinates are *momentum* and *position*.

match Johnson's experimental result (Johnson, 1928) that voltage fluctuations were proportional to resistance and to temperature—relates the effect to circuit parameters of resistance, inductance and capacitance which may be treated on a 'black box' basis. For this one does not need to know the internal mechanism of the circuit components, but one still needs to know their temperature (or in some cases, effective temperature). This is based on the fact that equipartition is applicable to any type or size of body or system though it was first used in the kinetic theory of gases, for example in connection with the ratio of specific heats at constant pressure and at constant volume. Equipartition is independent of the specific type of energy, mechanism of storage and mechanisms of exchange of energy. (There must be some means of exchanging energy with the system in question if its energy content is to be observed.) A particular historical application of equipartition, of which some trace may be seen in Nyquist's derivation of his formula for resistance noise, is Rayleigh's (1900) application to the standing-wave modes of black body radiation in an enclosure with reflecting walls. Since the number of such modes is inversely proportional to the wavelength, this predicted that the spectral energy density would increase without limit as the wavelength decreased, a result known as 'the ultra-violet catastrophe'. This difficulty was eliminated by Planck's introduction of the quantum which would modify the spectrum at high enough frequency. The quantum correction in electrical applications was of only theoretical interest in 1928 but in the 1980's the effect might just be detectable at gigahertz frequencies combined with cryogenic temperatures.

Gas molecules are invisible but the generality of equipartition can be demonstrated by studying the interaction of molecules with a more massive object such as a small mirror on a delicate suspension: angular movements of the mirror are brought about by random fluctuations in the numbers and momenta of air molecules bombarding the mirror on either side of its central axis of rotation. Kappler (1931) set up a torsion pendulum consisting of a mirror of one or two square millimetres area and some tenths of a micrometre thick on a quartz fibre some centimetres long. From a photographic record of the random angular deflections, reproduced in Fig. 1.3(a), he was able to deduce Boltzmann's constant and hence obtain a value for Avogadro's number of $6.059 \times 10^{23} \pm 1\%$ which differs by only 0.6% from the now accepted value of 6.022×10^{23}. The torsion pendulum has two components of energy, the kinetic energy of the moving mass and the elastic energy in the suspension which provides the restoring

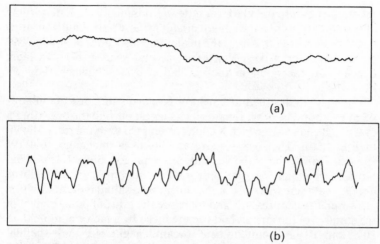

Fig. 1.3 Angular fluctuations of small torsion pendulum (from Kappler, 1931): (a) at atmospheric pressure and temperature 15°C, (b) at 4×10^{-3} mm mercury and temperature 10°C

force, so that the pendulum has two packets of equipartition energy, corresponding to two degrees of freedom.

'Degree of freedom' is a difficult concept to define, but in outline the number of degrees of freedom of an object or system is equal to the number of independent variables of which the instantaneous values must be specified in order to define its state. The familiar example from the kinetic theory of gases is that a monatomic ('billiard ball') molecule has three degrees of freedom, corresponding to the x, y and z components of momentum, but a diatomic ('dumb-bell') molecule has five, the two extra ones defining the rotation of the axis of the molecule in two planes (ignoring the spin about the axis which is unobservable and is therefore assumed not to contribute to the exchange of energy on collision between two molecules). The same principle applies to more complex systems, that the number of degrees of freedom is equal to the number of variables of which the values must be specified in order to define the state of the system. The question of spin of the diatomic molecule illustrates an important point—that one ignores degrees of freedom of which the energy cannot be observed. Thus in Kappler's experiments there were undoubtedly possibilities of flexural vibrations within the mirror but these would not affect the angular movement of the whole mirror, so Kappler could not observe them and was right to ignore them. In a parallel resistance-capacitance electrical circuit the voltage across the capacitance is sufficient to define also the current through the resistance, so the parallel RC (or series RL) circuit has only one degree of freedom. But the LC resonant circuit (or any harmonic oscillator) has two degrees of freedom: the instantaneous values of voltage and current must both be specified because the relation between them depends on the point in the oscillation cycle at which they are observed.

If one attributes the movement of Kappler's mirror to bombardment by air molecules it is natural to suggest putting the device in an evacuated enclosure, which Kappler attempted with the result shown in Fig. 1.3(b). The mean-square amplitude is unchanged but the *spectral distribution* of the fluctuation is changed. Since Kappler did not achieve a high vacuum, only 4×10^{-3} mm mercury, it might be argued that there was still bombardment by residual molecules of gas. However, the gas pressure, and therefore the rate of bombardment of the surfaces of the mirror, had been reduced by a factor of more than 10^5; and if gas bombardment became negligible, then thermal agitation in the quartz fibre would provide the exchange of energy. This thermal agitation would be associated with departure of the quartz fibre from perfect elasticity, a factor which would also provide the residual damping of the torsion pendulum. This illustrates the 'fluctuation-dissipation theorem' (Callen and Welton, 1951; Kubo, 1966), that any source of dissipation of energy will also be a source of fluctuation: this is familiar in the Nyquist formula associating fluctuations of voltage with resistance. If this were not so there could be continuous one-way transfer of energy, e.g. from the air to the torsion pendulum, which would contradict the second law of thermodynamics. Another important point is that equipartition applies to the *total* energy, regardless of the spectral distribution which differs between Fig. 1.3(a) and 1.3(b).

1.6 Nyquist

By applying the equipartition theorem to the modes of oscillation of standing waves on a loss-free transmission line (cf. Rayleigh's standing-wave modes in black-body radiation) Nyquist established both the magnitude of fluctuation to be associated with a resistance (now treated as a black-box circuit element) and its flat spectral distribution. The following treatment, which differs slightly from Nyquist's original treatment (Nyquist, 1928), has also been used by Bittel and Storm (1971). Consider first a low-loss transmission line of length l, open-circuited at each end (Fig. 1.4(a)). This will support a standing wave of every wavelength λ such that l is an integral multiple of $\lambda/2$. The number of such modes of standing wave within a narrow frequency band $d\nu$ is $2l\, d\nu/c$ where c is the velocity of propagation

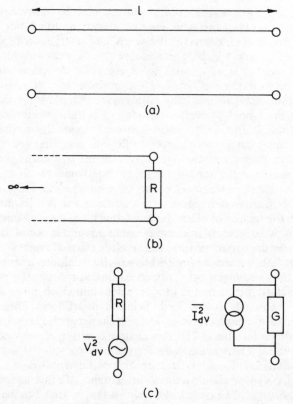

Fig. 1.4 Derivation of Nyquist's formula: (a) open-circuited low-loss transmission line of length l, (b) line of infinite length terminated by matched resistance R, (c) voltage and current equivalent circuits of resistance plus noise source

along the line. Each has energy kT (two degrees of freedom*) so on dividing by l the energy per unit length of line is $2kT\,\mathrm{d}v/c$. Now let l tend to infinity. While increasing the number of possible modes, this increases both the total energy and the length at the same time, leaving the energy per unit length unchanged. Each standing wave consists of a pair of equal waves travelling in opposite directions with velocity c, leading to a *power* flow kT in each direction. Next, let the line be cut in the neighbourhood of the observer, the right-hand half being discarded and the (infinite) left-hand half given a matching termination consisting of a resistor of value R equal to the characteristic impedance of the low-loss line, as shown in Fig. 1.4(b). The

---

* Nyquist called each mode a degree of freedom, but then attributed to each an energy kT.

resistor with its internal noise source may be represented by either of the two equivalent circuits shown in Fig. 1.4(c). If the resistor contains a generator $\bar{V}_{dv}^2 = 4RkTdv$ or $\bar{I}_{dv}^2 = 4GKTdv$, the maximum power which it can deliver to a matched load (the *available power*) is $kTdv$ and so is equal to the power flowing along the transmission line to the resistor in Fig. 1.4(b). The principle of detailed balancing (Bridgman, 1928) ensures that this equality must hold at every frequency: if it did not, the insertion of a suitable frequency filter between transmission line and resistor could cause more power to flow in one direction than in the other between two passive elements, leading to a contravention of the second law of thermodynamics.

At one time there was some misunderstanding about the noise and number of degrees of freedom of a long line such as a transatlantic cable. In the treatment of the 'Nyquist' line it has been assumed that there are two degrees of freedom for each possible mode of standing wave on the unterminated line and therefore many degrees of freedom for a long line. But *viewed from the terminals* the infinite or terminated line has only one degree of freedom because it appears purely resistive and therefore the current is uniquely determined when the voltage between terminals is specified. (This is apart from attenuation between remote parts of a lossy line and the accessible terminals: see discussion on formula (1.25).) It is again a question of noise *power* at the terminals versus noise *energy* in the line.

Moullin (1938) found difficulties with Nyquist's work. Firstly he asked why Nyquist should *assume* the existence of a fluctuation e.m.f. in the resistor R. The practical answer is that adjacent to Nyquist's paper in the publication was that by Johnson (1928) reporting his experimental observations* of such an e.m.f., but the theoretical answer is in the work reported in the first part of this chapter. Nyquist also postulated energy in a *loss-free* transmission line, which would seem to conflict with the fluctuation–dissipation theorem. The loss-free feature seems unnecessary because (1) any line with mismatched termination will have the same set of standing waves and (2) a damped mode of oscillation will still have the equipartition energy, as illustrated by Kappler's torsion pendulum. Therefore the line in Fig. 1.4 is here described as low-loss, not loss-free. (The loss must be vanishingly small if the characteristic impedance is to be approximated by a pure resistance.) Moullin also queries the point that the resistor will have some stray reactance and therefore will not match

---

* Johnson's resistors included copper and platinum films, a wire-wound half megohm resistor and four different electrolytes. He expressed his results in the form $V^2/R = 4kT(f_2 - f_1)$.

the line at high frequencies. Difficulties of matching can be avoided if one works in terms of available noise power

$$P_A = kT \, dv \tag{1.12}$$

which implies that any reactance is cancelled out by the inclusion of a conjugate reactance. This treatment of the transmission line needs to be modified in accordance with quantum theory when the frequency v is high enough and the temperature low enough to abrogate the condition $hv \ll kT$. It has often been said that the quantum correction leads to

$$\bar{P} = \left[\frac{hv}{\exp(hv/kT) - 1} + \frac{1}{2} hv \right] dv \tag{1.13}$$

where the fraction is the usual quantum correction and the term $\frac{1}{2}hv$ is often called 'vacuum fluctuation' or 'zero-point energy' (the latter because it still exists at zero temperature). This extra term is based on the fact that if one takes eigenvalues of the appropriate differential equation the allowed energy states for a single isolated harmonic oscillator are found to be $(n + \frac{1}{2})hv$. Note that (1.13) applies to the *average* power. Since allowed changes of state are by whole quanta, the $\frac{1}{2}hv$ *cannot take part in any exchanges and is not to be included in the available* power. But other methods of quantisation avoid the term $\frac{1}{2}hv$ (Bogoliubov and Shirkov, 1959) and Kubo (1966) arrived at a formula for the average energy of a harmonic oscillator

$$\bar{P} = \frac{1}{2}hv \coth(hv/2kT) \tag{1.14}$$

Since $\coth u \simeq 1/u + u/3$ for small u, the right-hand sides of both (1.13)* and (1.14) can be approximated by

$$kT \left[1 + \frac{1}{12} \left(\frac{hv}{kT} \right)^2 \right]$$

1.7 The quantum limit on linear amplifiers

It has been argued that if one is working down to the quantum limit on signal structure, no linear amplifier can be noise-free (Heffner, 1962). The argument is based on the *uncertainty principle* that in any physical measurement the product of two conjugate quantities has a minimum value of uncertainty, i.e. there is a minimum r.m.s. value of spread of measured values about the average value. Conjugate quantities are such that their product has the dimensions of *action*, so that the energy of a system and the time at which the energy is

---

* See Appendix II, II.1, for approximation to Equation (1.13).

specified constitute one conjugate pair. For a radiation/electrical system this is translated into energy $nh\nu$ of n quanta and phase φ related to time by $\varphi = 2\pi\nu t$ so that the uncertainty principle becomes

$$\Delta n \, \Delta\varphi \geqslant \tfrac{1}{2} \qquad (1.15)$$

A perfect amplifier would multiply the number of quanta by G and shift the phase by a constant amount θ so that

$$n_0 = Gn_i = G(n + \Delta n) \qquad (1.16)$$

$$\varphi_0 = \varphi_i + \theta = \varphi + \Delta\varphi + \theta \qquad (1.17)$$

Then if the uncertainty principle is applied to the output, $G \, \Delta n \, \Delta\varphi \geqslant \tfrac{1}{2}$, it implies that uncertainty in the knowledge of the inputs is reduced by a factor G, which is impossible. Heffner deduces that the minimum value of the noise temperature T_n of a linear amplifier with additive white gaussian noise is

$$T_n = \left[\ln \frac{2 - 1/G}{1 - 1/G} \right]^{-1} \frac{h\nu}{k} \qquad (1.18)$$

and the minimum value of phase fluctuation occurring in the amplifier is given by

$$(\Delta\varphi_a)^2 = \frac{(G-1)h\nu B}{2P} \qquad (1.19)$$

where P is the signal power and B the bandwidth. If $G \to \infty$, one can transpose the T_n equation to read

$$h\nu = kT_n \ln 2 \qquad (1.20)$$

Thus noise energy $kT_n \ln 2$ is associated with a signal energy of one quantum $h\nu$. It is interesting to note that Szilard (1929) showed that a reduction in entropy of $k \ln 2$ was associated with the extraction of one bit of information from a binary system. It is reasonable that if $G = 1$ the amplifier need introduce no phase noise; and that otherwise the phase fluctuation is proportional to the bandwidth B and inversely proportional to the number of quanta in the input signal, $n_i = P/h\nu$. But the first stage in the detection of weak electromagnetic radiation is always a photoelectric converter, such as a photo-diode or a photo-conductive cell; and these are not ideal linear amplifiers but generate so much excess noise that the quantum or indeterminacy limits are irrelevant. Moreover, they are *square* law devices in terms of electric field, which might seem a more appropriate concept than photons when phase is under consideration. But it is not obvious that a square-law amplifier would circumvent the quantum limitation.

Berensee (1963) later commented that this limit was unrealistically low because one must treat the composite system comprising amplifier, loss mechanism and energy source. Unfortunately, the explicit equations which he gave were valid only for $t < 1$, though a correction factor for longer times was also given. He rightly regarded noise as non-stationary over such short intervals and worked in terms of ensemble averages, i.e. averages over many systems, instead of the time averages of stationary noise which are normally assumed for thermal noise and noise temperature. The Heffner minimum noise temperature for an ideal amplifier can be regarded as a lower limit which is unlikely to be attained and to which must be added the effect of noise from the signal source and from the amplifier load. As an example which might be relevant to the search for gravitational waves (Chapter 7), Equation (1.18) with a frequency of 1 kHz and a gain as low as 2 would indicate a noise temperature of about 4.4×10^{-8} K.

1.8 Generalised circuits

It is a simple extension of Nyquist's result to replace resistance and conductance by the real parts of impedance and admittance:

$$\overline{V_{\mathrm{d}v}^2} = 4kT\mathscr{R}(Z)\,\mathrm{d}v$$
$$\overline{I_{\mathrm{d}v}^2} = 4kT\mathscr{R}(Y)\,\mathrm{d}v \tag{1.21}$$

One can then express the relationship between Nyquist's values for the mean-square voltage or current within a narrow band of frequencies and the total (all-frequency) fluctuation indicated by equipartition by an integral:

$$\overline{V_{\mathrm{Tot}}^2} = 4kT \int_0^\infty \mathscr{R}(Z)\,\mathrm{d}v \tag{1.22}$$

Moullin and Ellis (1934) took the simple case of a parallel CR combination, for which $\mathscr{R}(Z) = R/(1 + \omega^2 C^2 R^2)$ and direct integration leads to $\overline{V_{\mathrm{Tot}}^2} = kT/C$ which accords with the equipartition value of energy in the capacitor having one degree of freedom, $\frac{1}{2}C\overline{V_{\mathrm{Tot}}^2} = \frac{1}{2}kT$.

This result can be generalised with the aid of the technique of contour integration which was used by Bode (1945) to derive a number of circuit theorems (Bell, 1953). If a circuit contains several reactances, each will contain its own equipartition value of energy. But, subject to the condition that all the component elements in the circuit are passive, linear and lumped, if the whole can be treated on a 'black box' basis as a two-terminal circuit, then it will have fluctuations at its terminals corresponding to only one degree of freedom. As a two-terminal network it must be minimum-phase and therefore amenable to contour-integration methods. Let its impedance be $Z = A + jB$ and assume that it is analytic at infinite frequency, which it will be for any finite number of component lumped elements, so that for large ω it can be written in the form

$$Z = A_\infty + jB_\infty/\omega + A_1/\omega^2 + jB_2/\omega^3 + \cdots \qquad (1.23)$$

(The real part of impedance is always an even function of frequency and the imaginary part odd.) Then integration round a closed contour consisting of the axis of real frequency and a semicircle at infinity shows that

$$\int_0^\infty (A - A_\infty)\, d\omega = -(\pi/2)B_\infty$$

(The left-hand side is half the integral along the real frequency axis and the right-hand side is half the integral round the semicircle.) Most two-terminal circuits will reduce to a stray capacitance C' at sufficiently high frequency so that $B_\infty = -1/C'$ and A_∞ is zero. Changing the integration from angular frequency to cyclic frequency, the integral then becomes

$$\int_0^\infty A\, dv = 1/4C' \qquad (1.24)$$

On multiplying through by $4kT$ this becomes

$$4kT \int_0^\infty \mathscr{R}(Z)\, dv = kT/C'$$

which results in the equipartition value of energy stored in the stray capacitance, $\frac{1}{2}C'V^2 = \frac{1}{2}kT$. If the circuit reduces to a residual inductance as frequency tends to infinity, a similar argument in terms of real part of *admittance* leads to $\frac{1}{2}LI^2 = \frac{1}{2}kT$. Note that either impedance or admittance must be chosen, as appropriate, so that the series representation at infinity may be in terms of inverse powers of ω as in (1.17). This completes one circle of argument, in which it has been shown from equipartition theory that any continuous-spectrum two-terminal circuit has average available noise power kT per unit bandwidth and then by integration of $\mathscr{R}(Z)$ over all frequencies that this is equivalent to the stray reactance of the circuit having average stored energy equal to $\frac{1}{2}kT$, the equipartition value for one degree of freedom. (The conversion between *energy* and *power* is brought about by the multiplication by frequency in the integration.)

A further practical development within the framework of classical circuit theory is the extension of the formulae by F. C. Williams (1937) to cover complex networks (still two-terminal, linear, lumped-

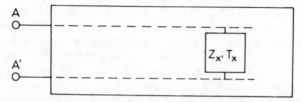

Fig. 1.5 General lumped network, with component elements at different temperatures

element circuits) in which the various elements are at different temperatures. For the general case represented schematically in Fig. 1.5, Williams proposed the following notation. Let Z_{xA} be the ratio of a voltage injected in series with Z_x to the current which it would cause to flow in a short-circuit between the terminals AA' and let $Z_{AA'}$ be the impedance seen on looking into the terminals AA'. Since with each element such as Z_x there is to be associated an equivalent voltage generator defined by $\overline{V^2} = 4kT\mathscr{R}(Z_x)\,dv$, the resultant mean-square fluctuation voltage which will be found at the terminals AA' on open circuit is

$$\overline{V_{dv}^2} = 4kT\,dv Z_{AA'}^2 \sum \frac{\mathscr{R}(Z_x)T_x}{Z_{Ax}^2} \tag{1.25}$$

The corresponding formula for mean-square current in terms of admittances is obtained by writing Y in place of Z. In practice, however, there may be non-thermal noise sources, e.g. in transistor equivalent circuits, and it is more usual to include the various noise generators in the complete equivalent circuit and find the total noise at any desired pair of terminals by *ad hoc* circuit analysis.

1.9 The temperature of radiation resistance

There was at one time reluctance to admit that the radiation resistance of an aerial could be the seat of thermal (Johnson) noise on the grounds that there was no corresponding mechanism of dissipation of electron energy *within* the aerial and therefore no means of assigning a temperature to the radiation resistance. But the discussion of equipartition emphasised that the thermal energy content of a system in equilibrium is independent of the mechanism by which it exchanges energy; and, moreover, the mechanism of exchange can be specified in the case of radiation resistance. Burgess (1941) pointed out that this mechanism was the exchange of radiation between an aerial and its surroundings: if the aerial were within a conductive enclosure the temperature of its radiation resistance would be that of its enclosure, but in free space it would be zero, i.e. the radiation resistance would not be a source of thermal noise if it radiated energy out but received nothing back. (It is now known (Penzias and Wilson, 1965) that outer space is not at zero temperature but contains a universal density of black-body radiation corresponding to a temperature of a few Kelvins. It has been hypothesised (Dicke *et al.*, 1965) that this is the attenuated remnant of the intense radiation which would have accompanied a 'big bang' creation of the universe.) In practice, the aerial is not specifically enclosed but the thermal radiation received is related to the attenuation of transmitted radiation; and if the aerial is directional and steerable, the effective temperature varies with the direction in which it is pointing. This is now a familiar experimental fact in communication with artificial satellites. If the communication path is near the zenith, so that there is little atmospheric attenuation, the temperature of the radiation resistance approaches that of outer space; but if the path is near the horizon, so that there is much atmospheric attenuation, the temperature of the radiation resistance then has the much higher value of the temperature of the earth's atmosphere.

Any internal loss resistance arising in the aerial's conductors or insulators naturally contributes additional thermal noise corresponding to the temperature of the structure.

1.10 Experimental evidence

The outcome of all the work outlined above is that, provided one assumes equipartition to apply to the macroscopic degrees of freedom of an electric circuit, it can be shown that all circuits obey the same law which can be expressed *either* in terms of mean available noise power for every frequency interval dv as in Equation (1.12), *or* in terms of mean energy $\frac{1}{2}kT$ per degree of freedom. The two expressions are mathematically equivalent for passive circuits consisting of lumped linear impedances; and for a transmission line the energy is kT per wavelength so that *energy* kT/λ is related to *power* $kT\,dv$ through the velocity of propagation along the line. The assumption that equipartition applies to electric circuits can be confirmed experimentally by calculating the value of Boltzmann's constant k from measurements of circuit noise as shown in Table 1.2. (The accepted value is now 1.38×10^{-23} J/K.)

1.11 Noise thermometry

The application of equipartition to the degrees of freedom of a mechanical system was confirmed by Kappler's measurements on

Table 1.2

Author	Value of k $\times 10^{-23}$ J/K	Estimated accuracy
Johnson, 1928	1.27	±13%
Ellis and Moullin, 1932	1.36	±1.4%

a torsion pendulum, as described above: he found $k = 1.372 \times 10^{-23}$ J/K. The idea of thermal noise in electrical circuits is now so well established that the 'noise thermometer' is an alternative to the gas thermometer for the establishment of the absolute (or Kelvin) scale of temperature. This aspect has been surveyed by Blalock and Shepard (1981), who report that an accuracy of better than 10 mK has been achieved at 90 K and an accuracy of 0.35 mK at 4 K. The noise voltage $\overline{V^2} = 4RkT\,dv$ can be measured (or compared with that of a reference resistor) by means of an amplifying system with a high input resistance or the corresponding noise current can be measured by a system with low input resistance. With the use of low-noise pre-amplifiers the noise contributed by the amplifying system is not usually important; but if necessary it can be eliminated by the correlation technique, in which the noise to be measured is applied to two similar but independent amplifiers and their outputs are cross-correlated. If the amplifier noise is random and the two amplifiers are independent, the cross-correlation of this noise is zero; but the noise to be measured is applied equally to both amplifiers and the corresponding outputs are fully correlated. The measuring apparatus usually employs a frequency band high enough to avoid any $1/f$ noise (e.g. 10 kHz to 200 kHz) and a moderately high frequency also facilitates the process of averaging the fluctuating voltage. Measurement in terms of either $\overline{V^2}$ or $\overline{I^2}$ as open-circuit or short-circuit values, respectively, requires a knowledge of the value of R (Fig. 1.4). But the product $\overline{V^2}\overline{I^2}$, a squared power, has the value $(4kT\,dv)^2$ from which it is said that the virtual power VI is $4kT\,dv$. This appears to conflict with Equation (1.12) which states that the (maximum) available power is $kT\,dv$. The explanation is that $\sqrt{\overline{V^2}}\sqrt{\overline{I^2}}$ is not the same as VI would be if both were measured simultaneously because V and I are negatively correlated if they both originate from a source of finite internal resistance. V and I are here defined as the open-circuit voltage and short-circuit current, i.e. it is impossible for both to occur simultaneously and VI is 'virtual' power. The measuring apparatus must be switched between the two modes, measuring V and I separately; and provided that the noise is stationary, the measured value of V is the same as it would have been if it could have been measured independently at the time when I was being measured and vice versa.

1.12 Thermal noise in non-ohmic devices in thermal equilibrium

The discussion in Sections 1.3 to 1.7 was limited to two topics: (1) the thermal noise generated in a homogeneous conductor (or semi-conductor) in thermal equilibrium with its surroundings was deduced from the mechanism of conduction; and (2) the thermal noise of a passive linear device was deduced on a 'black box' basis, after Nyquist. The first extension is to a two-terminal passive device which is non-linear but in thermal equilibrium. It is tempting to say that if the non-linearity, or the noise voltage, were small enough to allow a linear approximation to the V/I characteristic over the range of voltage which covers most of the noise (there are in principle rare noise peaks of unlimited amplitude), then one could immediately use the Nyquist argument of energy balance to conclude that the approximately linear device must be the source of a noise voltage

$$\overline{v_n^2} = 4kTB(\mathrm{d}V/\mathrm{d}I)$$

where $\mathrm{d}V/\mathrm{d}I$ represents the linear approximation. But if one expects to use the device in order to rectify signals down to the noise level, the linear approximation is clearly inapplicable and in Fig. 1.6 the noise associated with the non-linear device is shown generally as

$$\overline{v_n^2} = 4\mathrm{F}(V, I)kTB \tag{1.26}$$

where the function $\mathrm{F}(V, I)$ depends on the nature of the non-linearity

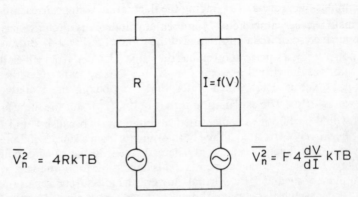

Fig. 1.6 Thermodynamic equilibrium between a passive non-linear device and a linear resistor

$f(\cdot V)$. A perfect rectifier, presenting a constant resistance dV/dI to voltages in one direction but infinite resistance in the opposite direction would pass only half current in response to an applied voltage having zero mean. It would therefore achieve energy balance with

$$F(V, I) = \tfrac{1}{2} \quad \text{and} \quad \overline{v_n^2} = 2kTB(dV/dI)$$

The practical approximation to the perfect rectifier is the Schottky diode which has an abrupt junction between metal and semiconductor (and, as such, is the successor to the catswhisker on a crystal).

In fact, one need not use the crude approximation of ohmic conductivity in one direction and zero in the other. Neudeck (1970) and Neudeck *et al.* (1972) analysed the noise in Schottky diodes in terms of the conduction characteristic

$$I = I_0[\exp(qV/\eta kT) - 1]$$

where $\eta \simeq 1$ is the 'ideality factor'. They derived an equivalent temperature of $\eta T/2$, which corresponds to $F = \tfrac{1}{2}$ in (1.26) if $\eta = 1$. They found experimentally that for six different units the ideality factor ranged from 1.05 to 1.4.

The whole topic has been reviewed from an engineering viewpoint by Gupta (1982), whose paper includes a full list of references to relevant work. From thermodynamic consideration he arrives at a general formula for the mean square open-circuit noise voltage of any device which is in thermal equilibrium with its surroundings:

$$\overline{v_n^2} = 4kTB\bar{P}_{ex}/\overline{y^2} \tag{1.27}$$

$\bar{P}_{ex}$ is the mean excess power dissipated in the device when there is applied a disturbance $y(t)$ of zero mean and mean-square value $\overline{y^2}$. In terms of the d.c. parameters V and I of the characteristic of the device, the noise voltage and current are

$$\overline{v_n^2} = 4kTB(dV/dI + \tfrac{1}{2}I\, d^2V/dI^2) \tag{1.28}$$

$$\overline{i_n^2} = 4kTB\left(dI/dV - \frac{1}{2}\frac{d^2I}{dV^2}\bigg/\frac{dI}{dV}\right) \tag{1.29}$$

It is particularly clear in (1.28) that the departure from the result for a linear device is jointly proportional to I, which includes any d.c. component resulting from rectification of noise, and to d^2V/dI^2 which is a measure of curvature of the I/V characteristic. For the usual type of curvature, e.g. in the power law

$$I = a_0 + a_1V + a_2V^2 + \cdots$$

the second derivative d^2V/dI^2 is negative so that both (1.28) and (1.29) indicate that there will be less noise from such a device than from a linear device. This accords with the result of Neudeck *et al.* for the Schottky diode as well as with the qualitative argument for an ideal rectifier.

1.13 Thermal noise in non-equilibrium conditions

Formulae (1.28) and (1.29) do not apply to devices which are not in thermal equilibrium with their surroundings, for example when there is a steady input of power to the device as when a diode or triode has a biasing voltage and current. This is in spite of the fact that it is now usual to describe noise due to fluctuations of velocity of electrons about their mean drift velocity as 'diffusion noise', which through the Einstein relation $D = \mu k T/e$ is equivalent to thermal noise. This has always seemed obvious (though not rigorously proved) provided the contribution to random velocity arising from scattering of the drift velocity remains small compared with the pre-existing random velocities. (If it does not, one has 'hot electrons': see Chapter 4.) As an extreme example, one assumes that the temperature of a quantity of gas can still be defined and measured when the gas is flowing through a pipe (Bell, 1938). This was, however, in the context of thermionic vacuum tubes in which the electron transit between cathode and anode is not influenced by any solid matter but only by space charge. The analogy is less exact in relation to solid state s.c.l. devices in which there is scattering by lattice collisions within the space charge region. In relation to space-charge-limited solid-state diodes the equivalence was formally noted by Van der Ziel and Van Vliet (1968). The Gupta relationship for non-linear devices does not apply here, since the system is not in thermal equilibrium but is being driven by the voltage applied to the diode.

As soon as one introduces an active device, such as a biased diode or a transistor, the whole argument in terms of thermodynamic equilibrium collapses so that one must examine the device in terms of internal mechanism rather than as a black box. A device which is non-linear is often also inhomogeneous in the sense that its material properties and electrical condition vary along the path of current through it—typically a path which is of either circular or rectangular cross-section and of constant area along its length. There are four methods of dealing with this situation, namely the salami method, the transfer impedance and impedance field methods and the generalisation in Langevin's transport equation. The last is considered the most rigorous; and much effort has been expended by various authors (e.g. Van Vliet *et al.*, 1975a) to show that the field impedance method is exactly equivalent to the Langevin method.

Nicolet *et al.* (1975) reviewed the noise in single and double injection devices and distinguished four sources of noise. One is $1/f$ noise (Chapter 2) which is significant only at low frequencies and so can usually be avoided by measuring at frequencies above a few kHz; injection noise, analogous to shot noise in the cathode emission of a thermionic device, is due to the random entry of electrons into the conduction path but is largely suppressed by space charge and is usually ignored in s.c.l. conditions; generation-recombination noise may be particularly important through recombination in a double injection device but has a characteristic time period and therefore frequency, so that it can be avoided by working at sufficiently high frequenty; and finally there is thermal or diffusion noise which is the main subject of investigation.

1.14 Salami methods

In the original salami method the device is imagined to be divided into a series of slices, each containing a noise voltage generator; and, since the noise sources are assumed to be uncorrelated, their combined effect is found by summing squared voltages. Van der Ziel (1966a) used this method to predict the noise current in a single-injection space-charge-limited diode. He assumed that each slice was a source of thermal noise in accordance with its resistance and temperature and found

$$\overline{i_{dv}^2} = 4kT(I_0/V_0)\,\mathrm{d}v = 8kT(\partial I/\partial V)\,\mathrm{d}v \qquad (1.30)$$

where I_0 and V_0 are the overall current and voltage. The second equality arises from the assumption of a square law, $\partial I/\partial V$ referring to values at the working point on the characteristic. This form of salami method was criticised by Thornber (1974) on the ground that it assumes that the noise in the several slices is uncorrelated, an assumption which is not valid in space-charge-limited conditions. An alternative suggestion had been that for an nth law device

$$S_V = 4kTn(\partial V/\partial I)$$

Since a space-charge-limited solid-state diode follows a square law (leaving aside disturbances due to traps) this leads to the same answer as Equation (1.30) for that particular case. However, Thornber regarded this as merely a coincidence and concluded that for any *space-charge-limited* solid-state device the voltage noise is

$$S_V = 4kT \cdot 2\,\mathrm{Re}[Z(\omega)] \qquad (1.31)$$

where $\mathrm{Re}[Z(\omega)]$ is the real part of the impedance of the device for small signals about the working point and at the working frequency. For comparatively low frequencies, where $Z(\omega)$ is real and equal to $\partial V/\partial I$, this reduces again to (1.30) for the s.c.l. diode. But it must be emphasised that (1.31) applies only to a space-charge-limited device and not to non-linear devices in general. Van Vliet et al. (1975a) proposed a modification of the salami method wherein a current generator in parallel with each slice replaces the voltage generator in series with it, but it is still the voltages across the slices which have to be added quadratically. In an ohmic circuit the two forms would be precisely equivalent; but in a non-linear device it is important that the current from the noise generator is divided between two paths, the approximately linear resistance of the slice and the non-linear resistance of the rest of the device. They stated (1975b) that the salami method was valid if the integral along the current path of a certain function of electric field, mobility and electron temperature and density was zero:

$$\int_0^L \mathrm{d}x[E_0(L) - 2E_0(x)]\mu(x)T_e(x)n_0(x) = 0 \qquad (1.32)$$

The modified salami method finds the contribution from each slice to the total device noise in terms of the voltage at the overall terminals resulting from a current at a particular point in the device, which immediately leads to the concept of transfer impedance (compare Equation (1.25), for transfer impedance in a network of lumped elements). When at the same time the slices are made infinitesimal so that the sum becomes an integral, we have what Van Vliet et al. (1975a) called 'the transfer impedance method' for finding the noise from a non-uniform device. There is a philosophical difficulty in making the slices vanishingly thin, since the continuity implied by integration ceases to apply to the physical situation if the thickness is less than either the atomic spacing or the mean free path of charge carriers. Probably the solution is to say that one can interpolate a smooth curve through any small-scale irregularities of the physical situation and it is this interpolated curve (which correctly represents the physical situation on the scale of practical interest) which is integrated. In the field-impedance method (see below) Shockley et al. (1966) avoided this difficulty by working in terms of the spectral intensity per unit volume of a continuous quantity (vector dipole current) arising from the motions of individual particles. They also propounded a 'Principle of Linearity for Significant Deviations' which was apparently intended to justify the use of large-scale concepts, such as impedance, for the results of random motions of individual charge carriers. Another example of this problem is examined in Appendix I on shot noise, namely consideration of the legitimacy of supposing that a charge of one electron on a capacitor decays exponentially. However, if this problem of microscopic lack of continuity can be ignored, the transfer-impedance method works on the basis that the relation between noise current intensity at a point and noise voltage at the terminals of a device can be represented by a transfer impedance $Z_T(\omega)$. In the general case of noise arising at a point, Z_T must be a tensor in order to accommodate multiple space-co-ordinate components of both current at the point and effect at the terminals; but in the case of axial symmetry, where variation is in one co-ordinate only, Z_T can be a single quantity and 'vector' only in the circuit sense of having 'real' and 'imaginary' parts.

1.15 The impedance field method

The impedance-field method due to Shockley et al. (1966) is more general than the transfer-impedance method in that it can deal with effects limited to any region in the semiconductor: it does not assume the continuity of electric current (including displacement current) which is axiomatic in a circuit treatment. The original treatment includes the relating of noise to the local diffusion coefficient D which in turn is related to thermal energy through the Einstein relationship $qD = \mu kT$. One then recovers the Nyquist relationship between noise and resistance, thus verifying the method. This relationship is applicable without qualification only if the scattering of electrons makes their random motion isotropic and capable of being super-imposed linearly on their drift velocity. The separation of drift and thermal velocities can be applied to active devices in which a (mean) current is flowing, but care is then needed in defining the effective temperature. (See 'hot electrons' in Chapter 4.)

The impedance-field technique finds the influence on the external circuit of a localised current anywhere within the device under examination. In the general case this is a three-dimensional problem which requires the use of vectors, as does any field problem in contrast to a circuit problem. The space between the electrodes is divided into small elements of volume $\Delta x\, \Delta y\, \Delta z$, and one such small volume is identified by the subscript α. If charges q_j move with (vector) velocity $\mathbf{u}_j$ in Δv_α, then a vector dipole current (the analogue of the frequently-used circuit current-element $i\, dl$) is defined as

$$\delta\dot{\mathbf{P}}_\alpha = q \sum_j u_j(t) \qquad (1.33)$$

Now let $Z_{N\alpha}$ (a scalar quantity) be the coefficient relating noise voltage at the terminal to current in Δv_α (for small signals it must be possible to specify a linear relationship):

$$\delta V_N = Z_{N\alpha}\,\delta I_\alpha$$

If the vector current in volume α is specified in terms of the vector $\mathbf{r}$ joining α to the terminal,

$$\delta V_N = \nabla Z_{nr}\,\delta \dot{P}_\alpha = \delta I_\alpha \nabla Z_{nr}\,\delta \mathbf{r} \tag{1.34}$$

where ∇ operating on the scalar Z_{nr} is the gradient operator and $\delta \dot{\mathbf{P}}_a$ of Equation (1.33) may also be written as $\delta I_\alpha\,\delta \mathbf{r}$. The quantity ∇Z_{nr} plays a part similar to that of the transfer impedance between a lumped element in a two-terminal network and the network terminals (Williams, 1937). Both I_α and Z_{nr} are in general functions of frequency and time, but this is independent of the space operator ∇ provided only that the space dimensions are small compared with the wavelength.

The elements of volume are considered large enough to be self-contained and uncorrelated as sources of noise—e.g. having dimensions larger than the mean free path. A conceptual difficulty is the transition from elements of δv which are large enough to be uncorrelated to an integral in which the element of integration is infinitesimal and therefore of dimensions smaller than the correlation distance. Shockley *et al.* provide a kind of 'smoothing' from the elementary current element $\delta \dot{\mathbf{P}}_a$ to a spectral density of noise $S(\delta V_N, \omega)$ which is uniformly distributed throughout the volume. Another point of view would be the interpolation of a smoothly varying function, as suggested above for the salami method. There is a good deal of mathematical manipulation, but the important results are as follows. If $u_j(t)$ is the x component of velocity of the jth carrier in one Δv and u_c^2 is *an average over all j* of the autocorrelation of u_j,

$$u_c^2(t'-t'') = \langle u_j(t')u_j(t'')\rangle_j \tag{1.35}$$

then the diffusion of u at frequency ω is defined as

$$D(u,\omega) = \mathrm{Re}\int_0^\infty e^{i\omega t}u_c^2(t)\,\mathrm{d}t \tag{1.36}$$

$D(u,0)$ is the usual diffusion constant applicable to a system in equilibrium with u being the x component of velocity due to Brownian motion. Setting $\omega = 0$ and noting that $u_c^2(t)$ is an average over j, the integral over all time can be replaced by an average over all j multiplied by the correlation time:

$$D(u,0) = \int_0^\infty u_c^2(t)\,\mathrm{d}t = u_c^2(0)\tau_c = \langle u_j^2\rangle_j \tau_c \tag{1.37}$$

τ_c may be equated to the mean free time of the carrier and $\frac{1}{2}mu_c^2 = \frac{1}{2}kT$ if u is the x component of thermal velocity. This leads to $D(u, 0) = \mu kT/q$. Returning to $S(\delta V_N, \omega)$, the result obtained after averaging is

$$S(\delta V_n, \omega) = \int |Z_{Nr}|^2 4q^2 Dn \; d(\text{vol}) \qquad (1.18)$$

where q is the charge on each of n carriers. As an example, for a homogeneous conductor

$$q^2 Dn = q^2 (\mu kT/q)n = kT\sigma \qquad (1.39)$$

where σ is the conductivity. Then from (1.38) and (1.39)

$$S(\delta V_N, \omega) = 4kT \int |Z_{rN}|^2 \sigma \; d(\text{vol})$$
$$= 4kTR \qquad (1.40)$$

since $Z_{Nr} = Z_{rN}$ for equilibrium systems.

Inter-valley scattering within the moving domain of the Gunn diode is the one clear case of noise requiring treatment by the impedance-field method.

1.16 Diffusion noise

Both μ and D can be measured directly in semiconductors by injecting a short pulse of minority carriers at one point and noting (1) the mean time of transit to another point and (2) the dispersion of the pulse. Alternatively, mobility is often specified as Hall mobility. The mobility in a material is usually quoted in practical units of $\text{cm}^2/\text{volt-second}$, but in Equation (1.39) the unit of conductivity must then be mho/cm, i.e. the reciprocal of resistivity in ohm-cm. The mobility in c.g.s. units is 300 times the mobility in $\text{cm}^2/\text{vole-second}$.

It is generally considered that the most fundamental and rigorous method of calculating noise is through the Langevin transport equation; and agreement with this is seen as proof of the legitimacy of other methods. For electric current density with transport by electrons the Langevin equation may be written:

$$J_n = q\mu_n E + qD_n \; \partial n/\partial x + H_1(x, t) \qquad (1.41)$$

(The subscript n indicates that the parameter is to have the value relevant to electrons.) The first term represents the drift under the applied field E while the second term represents the mean effect of diffusion down the concentration gradient, an effect which is important in semiconductor junction devices. The third term serves to represent any other forces which may be present, including noise sources; and the inclusion of the random noise turns (1.41) into a stoachastic differential equation which may be solved through the introduction of a Green's function. It is the equivalence of this function to the transfer impedance or the like which makes the transfer-impedance and field-impedance methods equivalent to the use of the Langevin equation. This leaves entirely open the specification of the mechanism of the noise, but the additive form of (1.41) implies that the noise is independent of the drift.

It must, however, be emphasised that Langevin's 1908 paper was concerned with Brownian motion and in effect with a particular application of the fluctuation–dissipation correspondence. Being concerned with comparatively gross particles suspended in a liquid, he assumed that the dissipation could be described by the Stokes law which applies to a sphere in a viscous fluid. He introduced a random force into his differential equation, but avoided any mathematical complexity by assuming the random force to have a mean value zero and a mean-square value known from equipartition. The first assumption is likely always to be valid but the second may not be true in active systems. It is then necessary to retain the stochastic term in the differential equation, leading to the more general form which is known as Langevin's equation. The Langevin equation may be elaborated by specifying cartesian components instead of assuming a one-dimensional conductor or by specifying various components of the stochastic contribution. But the two points remain that (1) it assumes that the stochastic component is additive to the steady components and (2) the nature of the stochastic component must be determined independently.

1.17 The small-signal diode

The discussion above has been relevant to devices for which the first term in the Langevin equation—representing movement due to an applied field—is dominant. These include devices of the FET type which are at present of great importance due to their compatibility with the more popular technologies of integrated circuits. But the first solid-state device to be the subject of full investigation was the p-n junction diode, which forms a basis for the examination also of the bipolar transistor. The static characteristic relating terminal current I_D of the diode to terminal voltage V is

$$I_D = I_S[\exp(qV/kT) - 1] \qquad (1.42)$$

where I_S is the reverse saturation current. Neglecting the thermal noise generated in any resistance in series with the junction (the base resistance of a transistor is the most usual case of concern), the spectral intensity of noise current associated with I_D may be written (Van der Ziel, 1970)

$$S_i(f) = 2q(I_D + 2I_S) \tag{1.43}$$

Since the diode conductance at low frequency is

$$g_0 = \frac{dI_D}{dV} \simeq \frac{q}{mkT} \exp(qV/mkT) = (q/mkT)(I_D + I_S) \tag{1.44}$$

(m is unity for low current and increases slowly with increasing current), then Equation (1.43) may be written

$$S_i(f) = 2mkTg_0(I_D + 2I_S)/(I_D + I_S) \tag{1.45}$$

For zero current this reduces to full thermal noise in g_0. But reference to (1.42) shows that the condition $I_D = 0$ requires $\exp(qV/kT) = 1$, i.e. the limit as $qV/kT \rightarrow 0$. In this case the diode will be passive and d^2I/dV^2 will be negligible so that Gupta's formula (1.29) will also reduce to $\overline{i_n^2} = 4kTBg$. At the other extreme, $qV \gg kT$ and therefore $I_D \gg I_S$, Equation (1.45) will tend to $S_i(f) = 2kTg$, in agreement with Neudeck (p. 23). If the junction width is greater than the diffusion length for carriers, the conductance at high frequencies is increased by back-diffusion of carriers so that (1.43) is modified to

$$S_i(f) = 2q(I_d + 2I_S) + 4kT(g - g_0) \tag{1.46}$$

where g is the high-frequency conductance.

This appears to be the sum of shot noise and thermal noise, using the standard formulae $S_i(f) = 2qI$ for shot noise and $S_i(f) = 4GkT$ for thermal noise, though there is no obvious reason why the shot noise at the terminals should be related to $I_D + 2I_S$ rather than to the terminal current I_D. In fact, formula (1.46) relates to a small-bias junction diode in which the current I_D results from diffusion; and detailed theory (Shockley, 1949) shows that I_D is the resultant of two much larger currents, of holes and electrons respectively, which reduce by recombination to the small difference current I_D. Recombination is thus a major factor in both static characteristic and noise in this situation. Van der Ziel (1955) proposed a circuit analogue of the situation in which current is 'lost' within the system (by recombination), namely a dissipative transmission line defined by the equations

$$\frac{\partial E}{\partial x} = -RI$$

$$\frac{\partial I}{\partial x} = -GE - C\frac{\partial E}{\partial t} \tag{1.47}$$

The attenuation of current in the transmission line through the leakage G is the analogue of loss of holes through recombination in the junction diode; and the full set of five analogies is

E: excess hole concentration

I: hole current

R: $1/qD_p$

G: q/τ_p

C: q

where the suffix p indicates parameters for holes and τ_p is the hole lifetime. The analogue of the characteristic impedance

$$Z_0 = R^{1/2}/(G + j\omega C)^{1/2}$$

of the transmission line is

$$Z_D = [q^2 D_p (1 + j\omega\tau_p)/\tau_p]^{-1/2} \tag{1.48}$$

Van der Ziel assumed that the corresponding resistive component of the impedance at the terminals of the device must show Johnson noise in accordance with Nyquist's formula. This implies that the recombination must generate white noise, at least under the condition of thermal equilibrium. (There is some experimental evidence of g–r noise centering on a particular frequency under conditions of substantial current flow.)

For this and other reasons, the transmission line analogy is no longer much used. Buckingham and Faulkner (1974) carried out a detailed analysis from first principles, considering every 'event' to be equivalent to a short pulse of current (an impulse or Dirac function), an 'event' being either the individual thermal motion (free flight) of a carrier between collisions or an individual recombination, but they appear not to have considered the possibility of a correlation between the times of recombination and subsequent release of the same carrier, i.e. trapping time. For the ideal junction diode they recovered formula (1.46); but if there was a depletion layer between p and n, recombination in a depletion layer would give rise to full shot noise in this recombination current. This effect is small in practice.

1.18 Other types of noise

There are other types of noise which are found only at low frequencies: the relevant frequency range varies between specimens, but is typically from about 1 kHz down to the lowest frequency which can be observed. This type of noise will be discussed in Chapter 2. There is also the question of charge carriers which are not in thermal equilibrium with the host lattice, i.e. 'hot electrons' (or hot holes). These will be discussed, together with avalanche phenomena, in Chapter 4.

REFERENCES

Bakker, C. J. and Heller, G. (1939). 'On the Brownian motion in electric resistances', *Physica* (The Hague), **6**, 262–274

Beaufoy, R. and Sparkes, J. J. (1957). 'The junction transistor as a charge-controlled device', *ATE J.*, **13**, 310–327

Bell, D. A. (1938). 'A theory of fluctuation noise', *J. IEE*, **82**, 522–536

Bell, D. A. (1953). 'Johnson noise and equipartition', *Proc. Phys. Soc. B*, **66**, 714–715

Bell, D. A. (1960). *Electrical Noise*, Van Nostrand; London

Berensee, R. M. (1963). 'The fundamental noise limit of linear amplifiers' (letter), *Proc. IRE*, **51**, 245

Bittel, H. and Storm, L. (1971). *Rauschen*, Springer-Verlag; Berlin

Blalock, T. V. and Shepard, R. L. (1981). 'Survey, applications, and prospects of Johnson noise thermometry', *Sixth International Conference on Noise in Physical Systems* (NBS Special Publication 614), 260–268

Bode, H. W. (1945). *Network Analysis and Feedback Amplifier Design*, Van Nostrand; New York

Bogoliubov, N. N. and Shirkov, D. V. (1959). *Introduction to the Theory of Quantised Fields*, Interscience Publishers Ltd; London

Bridgman, P. W. (1928). 'Note on the principle of detailed balancing', *Phys. Rev.*, **31**, 101–102

Brillouin, L. (1934). 'Fluctuations dans un conducteur', *Helv. Phys. Acta*, **7** (Suppl. 2), 46–67

Buckingham, M. J. and Faulkner, E. A. (1974). 'The theory of inherent noise in p-n junction diodes and bipolar transistors', *Radio Electron. Engr.*, **44**, 125–140

Burgess, R. E. (1941). 'Noise in receiving aerial systems', *Proc. Phys. Soc.*, **53**, 293–304

Burgess, R. E. (1959). 'Homophase and heterophase fluctuations in semiconducting crystals', *Discussions of Faraday Soc., No. 28, Crystal imperfections and the chemical reactivity of solids*, 151–158

Callen, H. B. and Welton, T. A. (1951). 'Irreversibility and generalized noise', *Phys. Rev.*, **83**, 34–40

Dicke, R. H., Peebles, P. J. E., Roll, P. J. and Wilkinson, D. T. (1965). 'Cosmic black-body radiation', *Astrophys. J.*, **142**, 414–419

Ehrenberg, W. (1958). *Electric Conduction in Semiconductors and Metals*, Clarendon Press; Oxford

Ellis, H. D. M. and Moullin, E. B. (1932). 'A measurement of Boltzmann's constant by means of the fluctuations of electron pressure in a conductor', *Proc. Camb. Phil. Soc.*, **28**, 386–402

Gupta, M. S. (1982). 'Thermal noise in nonlinear resistive devices and its circuit representation', *Proc. IEEE*, **70**, 788–804

Heffner, H. (1962). 'The fundamental noise limit of linear amplifiers', *Proc. IRE*, **50**, 1604–1608

Johnson, J. B. (1928). 'Thermal agitation of electricity in conductors', *Phys. Rev.*, **32**, 97–109

Kappler, E. (1931). 'Versuche zur Messung der Avogadro-Loschmidtschen Zahl aus der Brownsche Bewegung einer Drehwaage', *Ann. Phys.*, Germany, **11**, 233–256

Kubo, R. (1966). 'The fluctuation-dissipation theorem', *Rept. Prog. Phys.*, **29** (Pt. 1), 255–284

Langevin, P. (1908). 'Theory of Brownian motion', *C. R. Acad. Sci.*, **146**, 503–533

Lorentz, H. A. (1916). *The Theory of Electrons*, 2nd edn, Teubner; Leipzig

Moullin, E. B. (1938). *Spontaneous Fluctuations of Voltage*, Clarendon Press; Oxford

Moullin, E. B. and Ellis, H. D. M. (1934). 'The spontaneous background noise in amplifiers due to thermal agitation and shot effects', *J. Inst. Elect. Engr.*, **74**, 323–348

Neudeck, G. W. (1970). 'High frequency shot noise in Schottky barrier diodes', *Solid St. Electronics*, **13**, 1249–1256

Neudeck, G. W., Minniti, R. J. and Anderson, R. M. (1972). 'The ideality and the high frequency noise of Schottky-barrier-type diodes', *IEEE J. Solid St. Circuits*, **SC-7**, 89–90

Nicolet, M-A., Bilger, H. R. and Zijlstra, R. J. J. (1975). 'Noise in single and double injection currents in solids (I), *Phys. Status Solidi*, **70**, 9–45

Nyquist, H. (1928). 'Thermal agitation of electric charge in conductors', *Phys. Rev.*, **32**, 110–113

Penzias, A. A. and Wilson, R. W. (1965). 'A measurement of excess antenna temperature at 4080 Mc/s, *Astrophys. J.*, **142**, 419–421

Rayleigh, Lord (1900). 'Remarks upon the law of complete radiation', *Phil. Mag.*, **49**, 539–540

Shockley, W. (1949). 'The theory of p-n junctions in semiconductors and p-n junction transistors', *Bell Syst. Tech. J.*, **28**, 435–489

Shockley, W., Copeland, J. A. and James, R. P. (1966). 'The impedance field method of noise calculation in active semiconductor devices', in *Quantum Theory of Atoms, Molecules and the Solid State* (Ed. Per-Olov Löwdin), Academic Press; London, pp. 537–563

Szilard, L. (1929). 'On entropy reduction in a thermodynamic system through the intervention of an intelligent being' (German), *Z. Phys.*, **53**, 840–856

Thornber, K. K. (1974). 'Some consequences of spatial correlation on noise calculation', *Solid St. Electronics*, **17**, 95–97

Tolman, R. C. (1938). *The Principles of Statistical Mechanics*, Oxford University Press; London

Van der Ziel, A. (1955). 'Theory of shot noise in junction diodes and junction transistors', *Proc. IRE*, **43**, 1639–1646

Van der Ziel, A. (1966a). 'Thermal noise in space-charge-limited diodes', *Solid St. Electronics*, **9**, 899–900

Van der Ziel, A. (1966b). 'H.F. thermal noise in space-charge-limited solid state diodes', *Solid St. Electronics*, **9**, 1139–1140

Van der Ziel, A. (1970). 'Noise in solid-state devices and lasers', *Proc. IEEE*, **58**, 1178–1206

Van der Ziel, A. and Van Vliet, K. M. (1968). 'H.F. thermal noise in space-charge limited solid-state diodes—II', *Solid St. Electronics*, **11**, 508–509

Van Vliet, K. M., Friedman, A., Zijlstra, R. J. J., Gisolf, A. and Van der Ziel, A. (1975a). 'Noise in single injection diodes. I: A survey of methods', *J. Appl. Phys.*, **46**, 1804–1813

Van Vliet, K. M., Friedman, A., Zijlstra, R. J. J., Gisolf, A. and Van der Ziel, A. (1975b). 'Noise in single injection diodes. II: Applications, *J. Appl. Phys.*, **46**, 1814–1823

Williams, F. C. (1937). 'Thermal fluctuations in complex networks', *J. Inst. Elect. Engr.*, **81**, 751–760

Chapter 2

1/f Noise and Burst Noise

2.1 Introduction and characterisation of 1/f noise

The topic of this chapter will be dealt with under four headings: (1) characterisation of $1/f$ noise; (2) mathematical models; (3) experimental evidence and 'physical' theories; and (4) burst noise.

The defining characteristic of $1/f$ noise is that, in contrast to the flat or 'white' spectrum of thermal noise, its spectral intensity varies inversely as frequency. In the early days, when this was observed mainly in granular carbon devices (Frederick, 1931; Otto, 1935), it was known as contact noise; but its occurrence other than at contacts was demonstrated by Schönwald (1932), and after Montgomery (1949) most measurements were made on a four-terminal basis, with constant current, so that contact effects at the current terminals were eliminated. At this stage the phenomenon was often known as current noise because the noise having $1/f$ spectrum was found only during the passage of a steady current, being of intensity approximately proportional to the square of this current. It has since been shown not only that the $1/f$ phenomenon is a fluctuation in resistance but also that it is still present in the absence of current. Therefore 'current noise' is not a valid description. With the sensitivity of apparatus then available, Bittel and Scheidhauer (1956) were unable to detect $1/f$ noise in metallic wires, either resistance wire of 20 μm diameter or platinum Wollaston wire of 1.2 μm diameter. They concluded that if there were any current noise in metals, it must be associated with a volume of less than 0.7×10^{-25} mm^3 per Hz of detector bandwidth. At this stage it was tempting to call the phenomenon 'semiconductor noise', with the suggestion that its non-occurrence in metals implied that it was somehow associated with the variability of number of mobile electrons (or holes) to be expected in semiconductors but not in metals. This idea was eventually contradicted by Hooge and Hoppenbrouwers' (1969) observation of $1/f$ noise in a metallic gold film (having conductivity similar to that of the bulk metal); and the phenomenon is now known by the physically neutral term '$1/f$ noise'.

The main characteristic of $1/f$ noise, and the one which it is most difficult to accept, is that it has no lower limit of frequency. An exact f^{-1} law would lead both to an infinite value of spectral intensity at $f = 0$ and an infinite value of the total power integrated from $f = 0$ to $f = \infty$. Nevertheless it is true that no experiment has ever revealed a low-frequency limit beyond which the spectral intensity ceases to vary inversely as the frequency. The lowest frequency at which spectral measurements have so far been made is $10^{-6.3}$ Hz (Caloyannides, 1974)*. The term $1/f$ is used rather loosely for all spectra which follow, without detectable lower limit, a law f^{-x} with x between about 0.7 and 1.3; but in Caloyannides' observations the decade of lowest frequency had x particularly close to unity. (The exact value of x is important in relation to some of the mathematical models.)

At first sight it is a serious difficulty that the total noise power would be infinite if the $1/f$ law were followed down to zero frequency. But with an exact f^{-1} law the total power in a range f_1 to f_2 is

$$P = C \int_{f_1}^{f_2} df/f = C \ln(f_2/f_1) \tag{2.1}$$

This result is commonly expressed in the form that the noise is a constant per decade; and the infinity as $f \to 0$ is logarithmic. Flinn (1968) pointed out that the estimated age of the universe is only 10^{17} seconds. He then took as an extreme upper limit a frequency of 10^{23} Hz, one period of which corresponds to the time taken by light to travel a distance equal to the classical radius of the electron. (This is much higher than the frequency of visible light, which is less than 10^{15} Hz, corresponding to the fact that the wavelength of visible light is much greater than atomic dimensions.) This gives a maximum imaginable frequency range of 40 decades and hence a maximum noise power 40 times that of one decade. Taking an experimental result from Brophy (1968), Flinn calculated a noise power over 40 decades of 3.5×10^{-7} W when there was a d.c. input of 0.39 W. In practice, the low-frequency limit must be raised several decades because no specimen is as old as the universe; and the high-frequency limit can probably be lowered because one would expect a cut-off (masked usually by thermal noise) at a frequency corresponding to the time of atomic events. However, this does not affect the conclusion from the extreme case, that, while the 'infinity problem' may be important to mathematical theories, it is not of great practical consequence.

One of the characteristics of $1/f$ noise which must be taken into

---

* It has been questioned whether Caloyannides observed 'true $1/f$' noise; but he certainly observed noise with no sign of a lower frequency limit to a slope which was approximately -1 throughout and very close to -1 for the lowest decade.

account in any proposed theoretical explanation is the wide range of materials in which it occurs: carbon granules, carbon composition resistors, cermet resistors, (pyrolytic) graphite films, mosaic metal films, metallic films, metal whiskers, single-crystal semiconductors, glasses and possibly electrolytes. Some of the apparently non-electrical phenomena which also show a $1/f$ fluctuation may in fact be of electrical origin. For example, if the $1/f$ fluctuations in the time-keeping of quartz clocks were due to the maintaining circuit, this would be basically an electrical phenomenon. But $1/f$ fluctuations are also found in a variety of physiological phenomena (Musha, 1981) and in phenomena in which no microscopic mechanism is plausible, such as the size distribution of cities, frequency of occurrence versus heights of floods on the River Nile, earthquakes and thunderstorms. There is also frequency of usage versus length of words (Zipf's law) and intensity of sound versus pitch in music (Voss and Clarke, 1978). Some examples have been discussed by Tandon and Bilger (1976) and by Machlup (1981). In connection with this type of phenomenon one should remember that anything which has a $1/f$ spectrum is scale-invariant in its 'waveform' and so accords with Mandelbrot's theory of fractals (see Section 2.2 of this chapter). But in this book we shall be concerned primarily with those phenomena which can be attributed to some mechanism of electrical conduction.

2.2 Mathematical models

The physical difficulty over infinite power at zero frequency was practically disposed of by Flinn (p. 36), but there are still mathematical difficulties in formulating a quantitative explanation of $1/f$ noise. It may first be remarked that any bounded function may be represented over a finite range which does not include an infinite number of discontinuities (e.g. a sample length of noise) by a Fourier series in which the terms tend to decrease in amplitude with increasing frequency at least as fast as their harmonic order. (This follows from the mean-value theorem.) Then the intensity, or squared amplitude, decreases as $1/f^2$ at sufficiently high frequency. This does not necessarily conflict with the experimental $1/f$ law, since such noise is usually submerged in thermal noise at the high-frequency end, so that one does not observe the cut-off law. The more serious difficulty is that the power spectrum of a stochastic phenomenon is usually obtained from its autocorrelation function $\varphi(\tau)$ via the Wiener–Khintchine transform

$$S(f) = 4 \int_0^\infty \varphi(\tau) \cos 2\pi f \tau \, d\tau \qquad (2.2)$$

This is a cosine transform which must necessarily produce an *even* function and could never produce the odd function $1/f$: consequently many of the formulae which have been proposed offer only an approximation to $1/f$ over a limited range.

The obvious mathematical device, of replacing f^{-1} by $(f^2 + a^2)^{-1/2}$ with a arbitrarily small, is unattractive because it implies a flattening of the spectrum at sufficiently low frequency, which has never been found.

At a time when $1/f$ noise was commonly thought to be associated with contacts (although Schönwald had published contrary evidence in 1932) it was proposed that the noise was due to the diffusion of impurities over the contact surfaces (Macfarlane, 1950; Richardson, 1950); but approximation to a $1/f$ law depended on choice of geometry of the system in which diffusion took place and was effective over only a small number of decades. Diffusion has been revived more recently in the form of thermal diffusion in connection with temperature fluctuations in thin metal films (Voss and Clarke, 1976). It may have limited application here; but, apart from the question of frequency range, there is the conclusive objection that it predicts correlation over a distance much greater than is usually found experimentally.

The most popular form is the 'spread of time constants'. Any set of events occurring at random with a common relaxation time τ will have a spectrum with intensity proportional to $\tau/(1 + \omega^2\tau^2)$ (a Lorentzian spectrum) which is flat for $\omega\tau \ll 1$ and varies inversely as the square of frequency for $\omega\tau \gg 1$; and by superimposing the spectra for various values of τ with appropriate intensities one can obtain any desired overall spectrum. Van der Ziel (1950) showed that if one assumes a continuum of time constants between τ_1 and τ_2 with each component weighted by $g(\tau) = [\tau \ln(\tau_2/\tau_1)]^{-1}$ the combined spectrum takes the form

$$S(\omega) = C \int_{\tau_1}^{\tau_2} \frac{\tau_g(\tau)\, d\tau}{1 + \omega^2\tau^2}$$

$$= \frac{C}{\omega} \frac{\tan^{-1}\omega\tau_2 - \tan^{-1}\omega\tau_1}{\ln(\tau_2/\tau_1)} \tag{2.3}$$

where C is an overall scaling factor. It is then easy to say that the $\tan^{-1}$ terms tend to $\pi/2$ and zero as $\omega\tau_2 \to \infty$ and $\omega\tau_1 \to 0$, leaving $1/\omega$ as the only frequency variable term. But when one plots out the numerical values of the numerator of (2.3) for various values of τ_2/τ_1, as shown in Fig. 2.1, it becomes apparent that the spread of time constants must cover twice as many decades as the required match to f^{-1}.

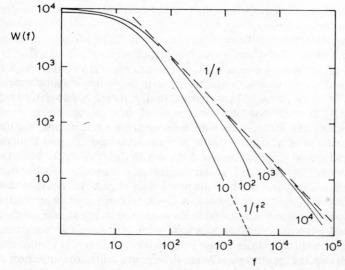

Fig. 2.1 Approximation to 1/f law by spread of time constants, according to Van der Ziel's formula

2.3 Two-stage or two-function models

It has also been proposed that $1/f$ noise could be produced by passing Poisson-generated noise through a suitable filter (Barnes and Allan, 1966; Radeka, 1969); but the physical nature of this filter remains hypothetical. Moreover the general idea of this type of two-stage process appears to have been ruled out by the experimental work of Voss (1978). With the concept that fluctuations occurred as deviations of a system from its norm, with subsequent relaxation towards the norm, Voss examined the question whether $1/f$ fluctuations occurred in systems that were linear in the sense that the shape of the relaxation curve was independent of the size of fluctuation. In mathematical terms, the system is linear if $(1/V_0)\langle V(t)\,|\,V(0) = V_0\rangle$ is independent of V_0, $t = 0$ being the time of occurrence of a particular fluctuation. Voss obtained smooth curves for $V(t)/V_0$, independent of V_0, by averaging over 32,768 samples of noise (32K in computer terminology) each of which was digitised at ± 1024 intervals from $t = 0$. A surprising feature was that the curve for negative time was the mirror image of that for positive time, i.e. the course of events leading up to V_0 was the time-reversal of that after V_0. This eliminates any causal process, such as applying pulses to a filter, which can only commence at $t = 0$; and it reinforces the view that $1/f$ noise is a property of the statistics of the collection of events rahter than of the characteristics of individual events.

An alternative to the two-stage process is the two-function process and a suggestion which might be compatible with the idea that $1/f$ noise is associated with mobility fluctuations is that a random walk in a random environment (rather than in an ordered environment) would produce an approximation to $1/f$ noise. A paper by Sinai (1982) showed that the movement of non-interacting particles between elastic collisions with fixed scatterers could look like the transitions of a random walk in a random medium and that the displacement of a particle would grow with time t as $(\ln t)^2$. Marinari et al. (1983) examined several cases of a random walk with a superimposed drift in which the probability of each step being to right or left was not fixed but was subject to some sort of spread of probability. If the two probabilities were fixed at $\frac{1}{2}, \frac{1}{2}$, at each step one would have the usual random walk with $1/f^2$ frequency spectrum; and if the probabilities might be only 0 or 1, chosen at random for each step, the spectrum would not diverge at $f = 0$. In the intermediate condition, where the probability distribution at each step had some sort of spread of values, multiplication of time by τ would correspond to multiplication of distances by $(\ln \tau)^2$ and the power spectrum would approach proportionality to $(\ln f)^4/f$. For the multidimensional case they supposed that the drift was due to the gradient of a potential V with a long-range correlation

$$\langle [V(x) - V(y)]^2 \rangle = |x - y|^{2\alpha}$$

and found a spectrum $S(f) \approx C(\ln f)^{2/\alpha}/f$. There are two problems here. Firstly, $(\ln f)^n/f$ is not a good approximation to $1/f$ for small f. Secondly, Marinari et al. postulate a superimposed drift; this would have been consistent with earlier ideas that $1/f$ noise is associated with current flow, but this is no longer accepted. There might, however, be scope for the work of Sinai, which did not involve a drift.

2.4 The Allan variance

The 'Allan variance', which was introduced by Barnes and Allan (1966), is a useful analytical tool which can be used on non-stationary stochastic series such as that representing $1/f$ noise. ($1/f$ implies that the longer you wait the greater will be the observed fluctuation; and this is clearly a non-stationary situation.) The Allan variance was first applied to the phase fluctuations of standard clocks and it was shown to be related to second differences in a series of numbers. Later, Van Vliet and Handel (1982) provided a mathematical proof of the relations (already stated by Barnes and Allan) between the time dependence of the Allan variance and the spectrum of the phenomenon to which it relates. The Allan variance may be defined as

$$(\sigma_n^T)^2 = \tfrac{1}{2}\big\langle (M_T^{(1)} - M_T^{(2)})^2 \big\rangle \tag{2.4}$$

where $M_T^{(1)}$ and $M_T^{(2)}$ are the time integrals of the stochastic function over the intervals $(t, t+T)$ and $(t+T, t+2T)$ respectively and the pointed brackets indicate an average over a sufficient number of pairs of intervals. The characteristic to be observed is the variation of the Allan variance with T, a variation which for $1/f$ noise approaches proportionality to T^2 as $T \rightarrow \infty$: if the spectrum is $C/|f|$, the Allan variance tends to $2CT^2 \ln 2$. (For Poisson shot noise, with a white spectrum of intensity m, it would be mT.)

2.5 Fractional order integration

A theory due to Mandelbrot that $1/f$ fluctuations represent fractional order Brownian motions is interesting because its possible application is very general. One expression of this theory is that, while integrating white noise once produces noise with a $1/f^2$ intensity spectrum, half-order integration will lead to $1/f$ (Mandelbrot and Ness, 1968)*. Mandelbrot further asserts that nature in general is 'fractal' (Mandelbrot, 1977), so that his approach covers also the non-electrical occurrences of $1/f$ fluctuations.

The mathematical and physical implications of half-order integration of white noise have been made clear by Press (1978). It turns out that it is mathematically equivalent to the convolution of white noise with a function G which can be written as a general linear combination of two component functions, one 'advanced' and the other 'retarded':

$$G_+(t-t_0) = \text{const.}(t-t_0)^{-1/2} \qquad \left.\begin{matrix} t > t_0 \\ t \leqslant t_0 \end{matrix}\right\} \quad \text{(a)}$$
$$\qquad\qquad = 0 \qquad\qquad$$

$$G_-(t_0-t) = 0 = \text{const.}(t_0-t)^{-1/2} \qquad \left.\begin{matrix} t \geqslant t_0 \\ t < t_0 \end{matrix}\right\} \quad \text{(b)}$$

If this convolution were to be effected by applying white noise to a physically realisable filter, G_- would have to be zero (no effect before the cause) and G_+ would have to be so valued as to account for the whole phenomenon. Press also points out that if it is to be regarded as a re-combination process, $dN/dt = \text{const.} \, N^n$ where n is the number of of entities involved in a re-combination, the value $n = 3$

---

* Half-order integration is an established mathematical concept. See R. Courant, *Differential and Integral Calculus*, Vol. 2, Chapter 4, Section 7 (London, Blackie 1936 and various reprints).

from a hypothetical three-body re-combination would give $N = $ const.$(t - t_0)^{-1/2}$ which corresponds to (a) above and therefore represents $1/f$ noise. Unfortunately the work of Voss (1978) showing experimentally that $1/f$ noise has time symmetry seems to rule out any mechanism of the dN/dt type—unless one can introduce a growth which on average follows exactly the same law as the decay. But it should be noted that Nelkin and Tremblay (1981) stated that Voss's test was a necessary but not sufficient test of time reversibility, particularly as it was based on averages, not on individual events.

2.6 Infra-red divergence

A quite different theory, applicable to electronic phenomena only, is due to Handel (1975) and claims that $1/f$ noise is associated with an essential infra-red divergence of quantum effects which may be classed generally as 'bremsstrahlung'. This term refers to radiation from a charge carrier which is subjected to negative acceleration; and if this radiation is equated with energy loss, it must always be found in association with resistance. Besides photons, the possible loss mechanisms* include 'transverse phonons, spin waves, configurational or correlated states, electron-hole pairs at the Fermi-surface of a metal, hydrodynamic excitations, etc., which have infra-red divergent coupling to the carrier' (Van Vliet *et al.*, 1981). The paper cited, entitled 'Superstatistical emission noise', examined the case of a Poisson process of which the mean rate was itself subject to fluctuation and applied it explicitly to the emission of electrons. If the fluctuation of the mean is also Poisson, the total noise is found to be shot noise plus modulation noise, the latter being of Lorentzian form (not $1/f$). The inclusion of transverse phonons, etc., makes it possible to include in the quantum theory the flow of current through a resistor; and the quantum interpretation of energy losses with appropriate approximation, leads to a noise term of $1/f$ form. It is not clear how it would account for the $1/f$ fluctuation which has been found in thermal noise, in the absence of a current; and work on the infra-red divergence theory lacks quantitative comparison with detailed experimental evidence from practical devices.

The theory of infra-red divergence due to quantum effects has also been applied to dielectric loss which is found empirically to follow a law $\chi''(\omega) \propto \omega^{n-1}$ where $0 < n < 1$ (Ngai *et al.*, 1979). A parallel is drawn with the X-ray absorption edges of metals and the I-R divergence depends on the existence of many correlated states (see also Dissado

---

* Handel's theory has been rejected by Dutta and Horn (1981) on the ground that it is relevant only to a beam of electrons at zero temperature. This list of possible loss mechanisms is presumably an answer to that objection.

and Hill, 1979). The formula accounts for a *constant* power law between ω^0 and ω^{-1} but not for a steeper slope than ω^{-1}. In general, the law applies only *above* the frequency of maximum loss, but for hopping charges a law of approximately ω^{-1} (ω^{n-1} with $n < 0.3$) applies also at low frequencies.

2.7 Physical dimensions

A certain amount can be deduced from consideration of the physical dimensions of the quantities involved. Using the S.I. system of metre, kilogramme, second and ampere which is equivalent to the c.g.s. magnetic system except that the dimension of the magnetic permeability constant is absorbed in the ampere, the analysis is as follows. Let the mean square fluctuation in a narrow frequency band be given by

$$\overline{\delta V^2} = Ci^\alpha R^\beta f^\gamma v^\eta \, df \tag{2.5}$$

where C is a purely numerical constant and V is volume. The MKSA dimensions of the quantities in (2.5) are: potential $= ML^2T^{-2}A^{-1}$; current $= A$; resistance $= ML^2T^{-2}A^{-2}$; frequency $= T^{-1}$; volume $= L^3$. The dimension of mass enters on the left-hand side of (2.5) only through voltage or potential, which can be defined in terms of the mechanical force between charges, and on the right-hand side through resistance, which involves potential. It follows that $\beta = 2$, provided only that the fluctuation voltage is *some* function of resistance. The equation for T is $-4 = -2\beta - \gamma - 1$ so that $\gamma = -1$; for A, $-2 = \alpha - 2\beta$ so that $\alpha = 2$; and for L, $4 = 4 + 3\eta$ so that $\eta = 0$. The result $\alpha + \gamma = 1$, relating the current dependence to the frequency dependence, was established by Macfarlane (1950). The finding that $\eta = 0$ does not seem to accord with common experience that, other things being equal, a bulky resistor is less noisy than a small one of the same ohmic value. But the finding that $\eta = 0$ in (2.5) does not rule out dependence of noise on volume and geometry of the resistor: the difficulty is that one cannot change either of these two factors independently without also changing resistance. One could also put $\eta = -1$ if C were not purely numeric but included a factor proportional to volume; and such a factor could be V/N, the inverse of the numerical density of charge carriers (Bell, 1960). Cancelling out the volume, the noise would be inversely proportional to the number of charge carriers involved as in Hooge's empirical formula (see below), $\delta R^2/R^2 = (\alpha_h/Nf) \, df$. Hooge suggested that α_h was a universal constant, but this need not be so if C in formula (2.5) can have different numerical values for different materials.

To take account of resistance fluctuations in the absence of current, one writes

$$\overline{\delta V^2}/i^2 R^2 = \overline{\delta V^2}/V^2 = \overline{\delta R^2}/R^2 = Cf^\gamma v^\eta \, \mathrm{d}f$$

The left-hand side of this is dimensionless so $\gamma = -1$ to match f with $\mathrm{d}f$, $\eta = 0$ and C contains $1/N$ so that, as before, $\overline{\delta R^2}/R^2 = C(1/Nf) \, \mathrm{d}f$.

Hooge (1969) used a dimensional argument to establish a relationship between his empirical formula for $1/f$ noise and the Nyquist formula for thermal noise, with the addition of a squared current and the elimination of the diffusion constant. (He introduced the latter via use of the Einstein equation $De = \mu kT$ to replace the kT factor in thermal noise and expressed both current and conductance in terms of atomic and geometric parameters.) But, in view of the subsequent discovery of $1/f$ noise in the absence of a current and the possibility that $1/f$ noise may be related to mobility, the elimination of diffusion is questionable. Incidentally, neither Hooge's formula nor (2.5) provides for a temperature effect. But it is arguable that the Boltzmann constant in kT has the dimensions of energy, so that temperature is a pure number and would not appear in dimensional analysis but could be added to Hooge's formula as a multiplying factor $f(T)$, the presumption being that the original formula applies to room-temperature noise.

2.8 Experimental evidence

It must now be stated explicitly that only $1/f$ noise in electrical conduction will be further considered; and that out of the enormous volume of experimental work which has been published during the past 30 years, only a few items will be cited in support of particular conclusions.

There is as yet no universally accepted theory of the mechanism of $1/f$ noise. The two major unanswered questions are:

(1) Is $1/f$ noise a surface effect or a volume effect?
(2) It appears to be a fluctuation in *resistance*, but is this due to a fluctuation in *number* or in *mobility* of charge carriers?

The supplementary questions are:

(3) What is the range of spatial correlation of $1/f$ noise and is it a truly scalar phenomenon?
(4) Is $1/f$ noise represented by a stochastic time series which is gaussian and stationary?
(5) Does $1/f$ noise vary with temperature in any consistent way?
(6) What is the status of the thermal-fluctuation theory of $1/f$ noise?

The question of surface versus volume affects the way in which one searches for a mechanism and in practice also the question of number versus mobility. With semiconductors it is almost universal experience that the magnitude of the 1/f noise can be greatly influenced by surface treatment, e.g. by suitable etching of the surface of the material. For example, Van de Voorde et al. (1979) found that 1/f noise in their specimens of InSb could be reduced by an order of magnitude by etching with CP4. At one time the 1/f noise in junction diodes was used as an inverse measure of quality and reliability: it was supposed that extra noise arose either from surface leakage round the edge of the junction (exacerbated by water vapour if the hermetic packaging were imperfect) or from imperfections in the structure of the junction. More recently, the incidence of 1/f noise in electric current and in light output has been proposed as a criterion of the reliability of diode lasers (Vandamme and Van Ruyven, 1983). In this case the relevant imperfections are presumably in the internal structure. There were two hypotheses about the mechanism by which surface imperfections generated noise. One was that the imperfection increased the speed of recombination of free charge carriers, i.e. reduced the life time. How this increased 1/f noise had still to be shown. The other hypothesis was that surface imperfections provided 'surface states' which were at energy levels different from those of the regular crystal lattice; and that these abnormal energy states constituted traps for electrons (holes), with an occupancy time proportional to the depth (in energy) of the traps. McWhorter's theory (1957) combined the idea of traps with the spread of time-constants described in Section 2.2 to account for an approximately 1/f spectrum. A reasonable judgment now is that surface states are *one* source of 1/f noise, but that there is also a source (or sources) distributed throughout the volume of a conductor.

The idea that 1/f noise is a volume, rather than surface, effect has been espoused by Hooge. In 1969 he showed that the published measurements of 1/f noise in many semiconductor specimens could be shown on logarithmic scales to be scattered around a trend line representing

$$\frac{\langle \delta V^2 \rangle}{V^2} = \frac{\alpha_0 \, \mathrm{d}f}{Nf} \qquad (2.6)$$

where N is the number of charge carriers involved and α_0 is a constant of magnitude approximately 2×10^{-3}. He subsequently modified his theory by postulating that only collisions between carriers and lattice would be effective in producing 1/f noise: scattering by impurities would not be. His modification of the theory then consists in replacing α_0 in (2.6) by

$$\alpha = (\mu/\mu_{\text{latt}})^2\alpha_0 \tag{2.7}$$

where μ is the observed mobility and μ_{latt} is the mobility which would be found if the scattering were due to the lattice only. Measurements on bismuth presented a difficulty to the original Hooge theory, since its density of mobile carriers is very much less than that in 'good' metals such as gold, silver, or copper but its $1/f$ noise is comparable. However, it has been claimed (Hooge *et al.*, 1979) that after allowing for scattering at surfaces results from a bismuth film are consistent with formula (2.7). The work of Dutta *et al.* (1977) on $1/f$ noise in copper whiskers also suggests a surface effect. In order to avoid substrate effects they used whiskers of copper about 3 μm in diameter and found $1/f$ noise levels about 2×10^3 higher than those calculated from Hooge's original formula for a bulk effect and 3×10^2 higher than that found in copper films of the same volume. This could be associated with the high surface/volume ratio, and the metallurgical structure, of whiskers; but perhaps it could also be argued that the effective mobility is affected by scattering of electrons from the surface, so that the modified Hooge formula (2.7) would still be valid.

Although it is customary to refer to lattice scattering, what is really meant is phonon scattering, for it is possible to change the *lattice* by melting a metal such as gallium without making much difference to the $1/f$ noise (Kedzia and Vandamme, 1978). It has also been shown that a grain boundary in a compound semiconductor added to $1/f$ noise (Hanafi and Van der Ziel, 1978), so that (2.7) is sometimes said to refer to elemental semiconductors. Fleetwood and Giordano (1983) measured the $1/f$ noise in evaporated films of gold, silver, gold–silver alloy, copper, tin and lead and in a sputtered film of platinum. All films had resistivities comparable with those of the corresponding bulk metals. They found a wide spread of noise levels for different specimens of the same metal but a *minimum* for each metal which showed a trend as between metals which was not related to the numbers of conduction electrons in the way required by Hooge's hypothesis. They proposed instead that the minimum noise spectral intensity for any metal could be represented by

$$S(V)_{\text{min}} = (\rho_0/\rho)(V^2/Nf^\alpha) \tag{2.8}$$

with $\rho_0 \simeq 6 \times 10^{-3}$ μΩ-cm and the index $\alpha = 1.08 \pm 0.08$ for $0.3 < f < 100$ Hz. Since ρ is proportional to $1/\mu$, the important difference between (2.8) and (2.7) is the use in (2.8) of the first power of ρ_0/ρ in contrast to the square of μ/μ_{latt} in (2.7).

Field-effect transistors are nowadays favourite test specimens; and those which are made by the MOS technique show evidence that much of the $1/f$ noise originates where the oxide insulation forms one boundary of the channel. It was long assumed that this noise was generated by the McWhorter mechanism, with electrons penetrating into traps at varying depths within the oxide. But Weissman and co-workers (Black *et al.*, 1983) found that noise was *increased* by etching away most of the oxide and therefore concluded that the noise must arise at the interface rather than within the oxide.

At one time it seemed plausible that the source of $1/f$ noise in the volume of a single-crystal semiconductor might be dislocations acting as 'internal surfaces'. However, this has been dismissed firstly by thorough measurements on deformed crystals (Stroeken and Kleinpenning, 1976) and finally by the detection of $1/f$ noise in (point contacts with) liquid metals which cannot contain any dislocations. Mercury at 300 K was investigated by Stroeken and Kleinpenning; and liquid gallium by Kedzia and Vandamme (1978).

One of the arguments in favour of a bulk effect has been the variation of noise with contact area in 'point' contacts. Assuming a hemispherical spread of current into the bulk material from the small ('point') area of contact, one can calculate both the spreading resistance and any noise produced in the bulk of the conductor. It is found that the mean square of such noise, expressed as $\langle (\Delta R/R^2) \rangle$, should vary as the cube of the resistance. (This circumvents the problem of measuring the area of contact.) Hoppenbrouwers and Hooge (1970) used crossed cylinders of various metals, and some contacts between hemispheres, and varied the contact pressure in order to vary the contact area. Some specimens came close to the R^3 law but others showed considerable variation. Weissman and co-workers (Black *et al.*, 1983) have suggested in an appendix to a paper that misleading results may occur due to the importance of the surface surrounding the contact, which will vary in radius as the contact area changes.

In the light of all the evidence one must assume that there are *both* surface and volume sources of $1/f$ noise, whatever may be the mechanism of the volume sources.

It seems probable that the form $\overline{\delta V^2} \propto i^2 R^2$ indicates a fluctuation in resistance

$$\frac{\overline{\delta V^2}}{V^2} = \frac{\overline{\delta R^2}}{R^2} = Cf^{-1} \, df \tag{2.9}$$

whether or not the constant C conforms to Hooge's hypothesis as set out in Equation (2.6). It has been confirmed by Jones and Francis (1975) that the observed fluctuation in voltage in carbon resistors is caused by a fluctuation in resistance. Their evidence depended on the

'1/Δf' noise, which is produced with a spectrum of the form $|f - f_0|^{-1}$ when a specimen is excited by an alternating current of frequency f_0; and, if both d.c. and a.c. excitation are applied simultaneously, both $1/f$ and $1/\Delta f$ spectra are produced. Jones and Francis found good correlation, e.g. $96 \pm 0.05\%$, between noise at f_1 and noise at $|f_0 - f_1|$. Failure to obtain similar results with a wide range of other types (unspecified) of resistor was attributed to temperature coefficient of resistance or non-linearity*. There was naturally no correlation between noise at different frequencies, at f_1 and $|f_2 - f_0|$. The lack of evidence from materials other than carbon is unfortunate: it leaves room for the suspicion that the results might apply only to hopping conduction. However, there is other evidence.

One of the fundamental questions has been whether $1/f$ noise is a phenomenon which exists spontaneously in the conductor and is *revealed* by the passage of current, or whether it is *caused* by the passage of current. If $1/f$ noise is a fluctuation in resistance, and this fluctuation exists in the absence of a steady current, then there should be a $1/f$ fluctuation in the $4RkT$ spectral intensity of the mean square voltage of Johnson noise. The normal interpretation of the Nyquist formula is that it averages the squared voltage over a time tending to infinity and so would average out any fluctuations in R and T. If one can take the average over a suitably short time (but long enough to give an average over the response time of the observing system, i.e. greater than the reciprocal bandwidth) this average should show any $1/f$ fluctuations in R. Such fluctuation is very small so that a specimen with a high value of $1/f$ noise is essential: Voss and Clarke (1976) used a film of indium with width reduced at one point to leave only a small bridge containing about 10^6 atoms; and with this they found a $1/f$ component of thermal noise. Beck and Spruit (1978) gave a detailed analysis of the time-constant requirements and reported finding a $1/f$ component in the thermal noise of carbon-paper resistors. It is thus seen that $1/f$ noise is something which occurs in the passive state of conductivity of the material.

The theory of these experiments has been considered in detail by Nelkin and Tremblay (1981) in their Section 4 and Appendix C; and they commented: 'These experiments show that $1/f$ noise is not

---

* There have been some reports of anomalous effects with pure a.c. excitation, such as the appearance of $1/f$ noise as well as $1/\Delta f$ noise; but this might be due to rectifying effects in the test resistor if it is of granular type involving numerous point contacts which might be partially rectifying. Certainly it has been established that there is correlation between the intensity of $1/f$ noise and a slight non-linearity in commercial resistors consisting of pyrolytic graphite film (Kirby, 1965). It is therefore probably correct to assume that $1/\Delta f$ noise is produced as a modulation effect between the alternating exciting current and the $1/f$ fluctuation of resistance.

caused by the applied voltage, but they do not show that it is a thermal equilibrium phenomenon.' They suggested that the most plausible physical mechanism of noise in metallic systems involves defect migration or other slow forms of structure fluctuation. Kogan and Nagaev (1982) have proposed a detailed mechanism for the generation of 1/f noise in metals by the dynamics of crystal structure defects, based on the experimental evidence of variation of noise intensity with temperature.

At this point one can note that logically the fluctuations in thermal noise might be caused by fluctuations in temperature, equally with fluctuations in resistance. If a body of heat capacity C_v is in contact with surroundings of very large (infinite) heat capacity at temperature T it will experience fluctuations in temperature of magnitude $\overline{\Delta T^2} = kT^2/C_v$.* If these fluctuations are to be detected through 'current noise' as a change in resistance, the resulting effect will depend on the temperature coefficient of resistance. This thermal fluctuation is significant if this temperature coefficient is very large, as at the edge of superconductivity (Ketchen and Clarke, 1978). At one time it was suggested that this might be the sole cause of 1/f noise, at least in thin metal films: but this has been ruled out in general because the spectrum of resistance fluctuations produced by temperature fluctuations is influenced by diffusion of heat through the specimen. This leads to prediction of a low-frequency cut-off which is not observed experimentally and to prediction of correlation of the noise over an appreciable distance, whereas the correlation distance of 1/f noise is known to be very small—probably less than one micrometre (Kleinpenning, 1977). It is now agreed that only in special cases do temperature fluctuations contribute significantly to 1/f noise.

2.9 Fluctuation in number or in mobility?

Given that 1/f noise is a fluctuation in resistance, the question whether it is due to fluctuations in number or in mobility of charge carriers is still disputed. Temperature variation of 1/f noise should give a clue, but the results differ so much from one experiment to another that no general law can be deduced. There is no evidence of an activation energy which might be expected to control numbers according to a law of the type $\exp(-\varepsilon/kT)$. On the other hand, mobility is expected to vary as $T^{-3/2}$ for phonon scattering, $T^{+3/2}$ for scattering by ionised impurities in heavily doped semiconductors or $T^{-1/2}$ in piezoelectric materials. With such a wide range of theoretical

* This is an equipartition type of thermodynamic law, independent of any electrical considerations: see, for example, Tolman, 1938, pp. 631–632.

values of temperature coefficient of mobility, coupled with variability of experimental data, one cannot make a good case for or against mobility fluctuation on grounds of temperature coefficient. Moreover, the laws just quoted apply to *mobility* and there is no guarantee that they would apply equally to *fluctuations in mobility*, i.e. that the fractional fluctuation is independent of temperature. (The same applies to number fluctuations, except that intuitively one expects the fractional fluctuation in number to be a constant function of average number, e.g. according to a Poisson law.)

Indirect evidence has therefore been adduced, including Hooge's μ/μ_{latt} hypothesis. Kleinpenning (1974) analysed the result of applying an additional e.m.f. to a thermojunction between semiconductors and showed that the resulting variation in $1/f$ noise fitted exactly the theoretical curve corresponding to the variable-mobility theory but did not accord with a variable-number theory. Hooge has shown that his $(\mu/\mu_{latt})^2$ theory accords with experimental results if proper allowance is made for scattering other than by the lattice, such as scattering by impurities in doped semiconductors and scattering at the surfaces of a thin bismuth film. For the latter, the correction was by a factor of up to 10 (Hooge *et al.*, 1979).

Support for a variable-number theory (and opposition to variable-mobility) has come from Weissman and co-workers, who have adopted the technique of observing the cross-correlation between noise parallel to the flow of current and noise at right angles to it (Weissman *et al.*, 1982). This firstly has the advantage of eliminating thermal noise, which is assumed to be purely random in direction and therefore will show no correlation between different directions, and secondly should show whether $1/f$ noise results from a purely scalar phenomenon, using comparison between the cross-correlation and the noise in individual directions. The general shape of wafer used is shown in Fig. 2.2, the perpendicular paths through the small central region being provided by the circuits AA' and BB'. An approximately constant current is driven through AA' by a battery plus series resistor R_s; and in order to maintain symmetry a second resistor R_s is connected across BB'. The four sectors are divided so as to provide separate current and potential terminals and the signals (noise) are taken from the pairs S_A, S_A' and S_B, S_B'. Measurements made (Black *et al.*, 1983) include (1) instantaneous isotropy of fluctuation, (2) ratio of magnitudes of fractional fluctuations in Hall coefficient and in resistivity and (3) correlation coefficient between fluctuations in Hall effect and fluctuations in resistivity. It was reported by Weissman *et al.* (1983) that a thick silicon wafer having no detectable non-phonon scattering and having homogeneous carrier concentration had a Hooge α value of 2×10^{-6}; and that a 71-μm silicon-on-sapphire

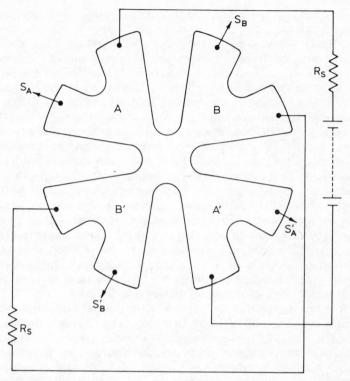

Fig. 2.2 Diagrammatic shape of wafer used by Weissman and co-workers for investigation of the characteristics of 1/f noise

wafer had a value of about 10^{-4}. The Hall noise was approximately as would be expected if the fluctuations were due to variations in number of charge-carriers. A complication is that any trapping mechanism will cause fluctuations in *both* number and mobility. For the trapping of a free carrier obviously represents a change in number; but its velocity is zero while it is in the trap, and hence its mobility, so trapping also produces a change in mobility. However, if the length of stay in the trap is longer than the transit time of the carrier through the specimen, as would usually be required to account for the low-frequency spectrum, only the long-term average mobility will vary on this account.

2.10 Is 1/*f* noise Gaussian?

There is then the question whether 1/*f* noise has a Gaussian distribution of amplitudes. It was shown at an early date that it was approximately Gaussian (Bell, 1955) but some of the later and more accurate measurements were less conclusive. However, Stoisiek and Wolf (1976) showed that the observed distribution of amplitudes from a noise spectrum of $1/f$ form would be affected by the ratio of upper to lower frequency of the pass band observed*. They observed both the spectra and the amplitude distributions in noise from a carbon film resistor, three metal film resistors and an *n-p-n* bipolar transistor; and they concluded firstly that the amplitude distribution of $1/f$ noise was the same as would result from putting Gaussian noise through a $1/f$ frequency filter and secondly that the statistical properties of $1/f$ noise in physical sources are fully consistent with the assumption of stationarity. (The assumption of stationarity is not consistent with the absence of a low-frequency cut-off; but this factor could not be effective when a low-frequency cut-off is imposed by the measuring apparatus. In experimental measurements there must always be such a cut-off frequency, even if it is only the reciprocal of the time of observation.) The fact that $1/f$ noise has a Gaussian distribution of amplitudes is not sufficient proof that it is a Gaussian *process*, to test for which one must examine the higher moments of the distribution (see, for example, Nelkin and Tremblay, 1981). The simplest such test is that the fourth moment should be three times the square of the second moment. In one direct test of this (Bell and Dissanayake, 1975) the fourth moment was in excess by 20–100% for a carbon resistor and both *p* and *n* silicon wafers, although Johnson noise produced the ratio of three within 2%. So far as the author is aware, the direct test has never been repeated, though there have been indirect tests (Stoisiek and Wolff, 1976) which conformed to Gaussian behaviour.

2.11 Amorphous solids

Most of the measurements of $1/f$ noise in solids have been on crystalline materials, either single-crystal semiconductors or polycrystalline metals, or else on mosaic forms of metal film in which conduction is undoubtedly controlled by hopping of charge carriers between islands. (Experiments on liquid metals are subject to reservations since they used point contacts of a different material; and experiments on electrolytic solutions are controversial.) Neudeck and Kriegel (1978) made systematic measurements on amorphous silicon films, using molecular-beam evaporation to prepare both films of amorphous pure silicon and films containing some hydrogen. None

---

* They used digital sampling and the relationship of their conclusions to the sampling theorem, or Nyquist rate of sampling, is discussed in Section 2.15.

of their spectra showed really good $1/f$ characteristics, and at a higher temperature, 150°C, the noise could be fitted to a generation-recombination process. Likewise, Kim and Van der Ziel (1980) could find no $1/f$ noise in hydrogenated amorphous silicon. Another example of absence of $1/f$ noise has been reported by Scofield and Webb (1983), who found the usual $1/f$ noise in gold, chromium, copper and nickel but in a niobium film 240-nm thick found noise at least two orders of magnitude below the prediction of the Hooge formula and having a different shape of spectrum. They suggested that this noise might have been due to diffusion of hydrogen through the film. Since glasses are amorphous, the glasses $V_2O_5:P_2O_5$ are amorphous semiconductors. The work of Sayer and Prasad (1979) is one of the few examples where an attempt has been made to estimate the number of charge carriers by a method independent of electrical conduction, namely electron spin resonance. They assumed that each vanadium ion which was converted from V^{5+} to V^{4+} contributed one charge carrier; and with this assumption they found that (1) the number of free carriers indicated by e.s.r. was approximately the same as the number predicted from the noise by Hooge's formula and (2) both the number of carriers and the noise power remained constant over the temperature range 77 K to 300 K, although the conductivity increased six-fold. The increase in conductivity would then be attributed to an increase in mobility which did not affect the noise. The conduction was said to take place by 'percolation'; and if this means that there was a meandering conduction path through a non-conducting matrix, it is questionable whether hopping may occur at some points. Conduction of this type may also be expected in the type of composite resistor known as cermet; and Van Calster *et al.* (1983) assumed from the temperature coefficient of resistance that their cermets (and an improved form known as a metanet) depended on variable hopping which gives rise to a broad spectrum of time constants. The present conclusion is that an elementary amorphous material (such as amorphous silicon) does not exhibit $1/f$ noise; but a glass or a cermet, the latter consisting of a mixture of conducting and insulating phases, does show $1/f$ noise. The implication of this and other evidence is that hopping conduction is a potent source of $1/f$ noise. It is not clear how this is related to surface effects.

2.12 Electrolytes

The existence of $1/f$ noise in electrolytic solutions is still controversial. In experiments on electrolytes the necessary high resistance/small volume of conductor is usually obtained from a small hole in a partition in the electrolytic cell and the question is whether this is to be regarded as a *hole*, analogous to the contact area of a 'point' contact between solids, with the $1/f$ noise arising in the spreading resistance in the volume surrounding the hole, or as a short *capillary* connecting two volumes of high conductivity and with $1/f$ noise arising within the capillary. The first published work (Hooge and Gaal, 1971) used an aperture of 10 μm diameter in a membrane 10 μm thick and regarded this as a *hole*. They found $1/f$ noise of very high value, with α proportional to concentration (on logarithmic scales) in the range 0.1 to 10 molar (gram-molecules per litre) with a number of different electrolytes. An opposite view was taken by de Goede *et al.* (1983), who used a hole of 1.70 μm radius (measured by electron microscope) and of length, i.e. thickness of the membrane, deduced from its resistance when filled with electrolyte, found to be 5.4 μm. They regarded this as a capillary and they considered only noise which occurred during bulk flow through it, flow due either to an electric field (electro-osmosis) or to a pressure difference. They found a spectrum of Lorentzian form which they attributed to concentration fluctuations in the electrolyte. The corner frequency *above* which the noise tended to fall as $1/f^2$ varied between 50 and 500 Hz and below the corner frequency the spectrum was flat at least down to 0.05 Hz. Earlier work (Van den Berg *et al.*, 1981) had concluded that if any $1/f$ noise were present it must have $α < 8 \times 10^{-3}$. Musha *et al.* (1983) found indications of both a Lorentzian type of spectrum which was independent of current density and a $1/f$ spectrum with intensity varying with current, but their results appear anomalous in two respects: (1) they found appreciable $1/f$ noise at zero current and (2) their $1/f$ noise increased with increasing current much faster than as the square of the current. However, the values of α which they deduced for varying concentrations followed the same logarithmic proportionality to concentration as had been found by Hooge and Gaal, though they extended the range to more dilute solutions. This consistent dependence on concentration deserves further investigation, although belief in $1/f$ noise in electrolytes in declining.

2.13 Burst noise

At one time some bodies subject to $1/f$ noise used also to exhibit an erratic noise known as 'burst noise' (for details see Bell, 1960). This sometimes took the form of sporadic peaks of noise, at intervals of the order of a millisecond; and when it occurred in carbon composition resistors it could be attributed to unstable contact between grains; while its occurrence in reverse-biased semiconductor diodes could be attributed to momentary breakdown with avalanche effects. At other times the effect in a semiconductor took the form of switching repeatedly between two otherwise stable states, again at intervals of the order of a millisecond. One suggested explanation for this was the presence of a metastable micro-plasma which controlled the switching between the two states. The effect is virtually unknown in modern devices; and one would discard any device which exhibited such a phenomenon.

2.14 Summary

In the absence of any completely acceptable theory of the mechanism of $1/f$ noise, one cannot predict but can only measure the noise produced by any particular device. But one can give some general advice on designing a system so as to minimise $1/f$ noise and some order-of-magnitude figures of the $1/f$ noise to be expected.

(1) The $1/f$ noise in *amplifiers* may be circumvented by using a carrier system, so that low frequencies may be excluded from the signal channel. It is of little value to use a.c. excitation of the device in which the $1/f$ noise arises, because this would merely result in the replacement of $1/f$ noise by $1/\Delta f$ noise which would appear in the signal channel.

(2) $1/f$ noise tends to follow the original Hooge rule of α/N so that, other things being equal, the smaller the device (the smaller the number N of charge carriers involved) the greater the $1/f$ noise. But the coefficient α of proportionality can vary widely—by a factor of 10^5 in an extreme case. It should be *less* than the primitive Hooge value of 2×10^{-3}.

(3) Whatever the intrinsic noise in the bulk of a semiconductor such as silicon or gallium arsenide, there is likely to be additional surface noise at external surfaces, internal discontinuities and interfaces such as that between silicon and silicon dioxide in a MOSFET. One therefore uses a suitable etch to 'clean' the external surface of a semiconductor and one seals it against moisture. Purity and homogeneity of the bulk of the semiconductor is also important: at one time a commercial company claimed reduction of $1/f$ noise in its products through 'perfect crystal technology'.

(4) If it is necessary to use any granular or otherwise inhomogeneous material, the 'grain' should be as fine as possible, as demonstrated by Van Calster *et al.* (1983) in their comparison of cermet and metanet resistors.

(5) The magnitude of $1/f$ noise power is most easily indicated in terms of the 'corner frequency' at which it is equal to Johnson noise and below which it grows inversely as the frequency. For bipolar

transistors this frequency may be 1 kHz, though it may sometimes be down to 100 Hz or less. For FETs the typical figure is 10 kHz, though in a GaAs MESFET the 1/ƒ noise has been measured up to 50 kHz (Suh *et al.*, 1981) and in some point-contact devices used in early radar reception the 1/ƒ noise was dominant up to MHz frequencies. The noise in non-metallic resistors depends so much on the internal structure that no generalisation is possible; but the 1/ƒ noise may be above Johnson noise up to 100 kHz. These figures refer to room temperature. There is some evidence that 1/ƒ noise in semiconductors may increase at low temperatures—say, below 125 K (Cox and Kandiah, 1981). Most devices would be damaged by an increase of 150°C above room temperature.

2.15 The sampling of 1/ƒ noise

When a quantity is estimated by taking a number of sample values, the variance of the group of samples is expected to be inversely proportional to the number of samples *provided they are independent*. Shannon's well-known sampling theorem has led us to regard it as standard practice to take two samples per cycle of the highest frequency involved in a low-pass system, or of the bandwidth in a band-pass system, independently of the relative amplitudes of the various frequencies involved, i.e. of the spectral shape within the passband. But this sampling theorem gives a *sufficient* number of samples for white noise rather than a *necessary* number. Qualitatively, the 1/ƒ noise cannot vary as rapidly as white noise and therefore the correlation time is longer: fewer *independent* samples can be obtained in the same time-bandwidth interval and therefore the variance is correspondingly greater. A quantitative analysis was presented by Stoisiek and Wolf (1976).

BIBLIOGRAPHIES

Very many papers on 1/ƒ noise have been published. The references here cited are those which the author considers to be either initial or important contributions to some aspect of the phenomenon. More complete bibliographies are to be found in the following reviews:

Dutta, P. and Horn, P. M. (1981). 'Low-frequency fluctuations in solids: 1/ƒ noise', *Rev. Mod. Phys.*, **53**, 497–516

Hooge, F. N., Kleinpenning, T. G. M. and Vandamme, L. K. J. (1981). 'Experimental studies on 1/ƒ noise', *Rep. Prog. Phys.*, **44**, 479–532

Van der Ziel, A. (1979). 'Flicker noise in electronic devices', *Adv. Electron. Elec. Phys.*, **49**, 225–297

Gupta, M. S. (1977). *Electrical Noise: Fundamentals and Sources* (Bibliography for Part III, Sections B and C), Wiley and IEEE Press; New York

REFERENCES

Barnes, J. A. and Allan, D. W. (1966). 'A statistical model of flicker noise', *Proc. IEEE*, **54**, 176–178

Beck, H. G. E. and Spruit, W. P. (1978). '1/f noise in the variance of Johnson noise', *J. Appl. Phys.*, **49**, 3384–3385

Bell, D. A. (1955). 'Distribution function of semiconductor noise', *Proc. Phys. Soc. B*, **68**, 690–691

Bell, D. A. (1960). *Electrical Noise*, Van Nostrand; London

Bell, D. A. and Dissanayake, S. P. B. (1975). 'Variance fluctuations of 1/f noise', *Electron. Lett.*, **11**, 274

Bittel, H. and Scheidhauer, K. (1956). 'On the question of noise in metallic conductors' (in German), *Z. Angew. Phys.*, **8**, 417–422

Black, R. D., Restle, P. J. and Weissman, R. B. (1983). 'Hall effect, anisotropy and temperature-dependence measurements of 1/f noise in silicon on sapphire', *Phys. Rev. B*, **28**, 1935–1943

Brophy, J. J. (1968). 'Statistics of 1/f noise', *Phys. Rev.*, **160**, 827–831

Caloyannides, M. A. (1974). 'Microcycle spectral estimates of 1/f noise in semiconductors', *J. Appl. Phys.*, **45**, 307–316

Cox, C. E. and Kandiah, K. (1981). 'The dependence of the low frequency noise of JFETs on device parameters and operating conditions', *Proceedings of 6th International Conference on Noise in Physical Systems* (Ed. P. H. E. Meijer, R. D. Mountain and R. J. Soulen, Jr), NBS Special Publication 614, pp. 71–74

de Goede, J., Roos, H., de Vos, A. and van den Berg, R. J. (1983). 'Electrical resistivity fluctuations in solutions of potassium chloride', *Noise in Physical Systems and 1/f Noise* (Proceedings of 7th International Conference on Noise in Physical Systems and 3rd International Conference on 1/f Noise. Ed. M. Savelli, G. Lecoy and J-P. Nougier), North-Holland; Amsterdam, pp. 393–396

Dissado, L. A. and Hill, R. M. (1979). 'Many-body interpretation of the dielectric response of solids', *IEE Conf. Rep. 177*, pp. 168–170

Dutta, P. and Horn, P. M. (1981). 'Low frequency fluctuations in solids: 1/f noise', *Rev. Mod. Phys.*, **53**, 497–516

Dutta, P., Eberhard, J. W. and Horn, P. M. (1977). 'Noise in copper whiskers', *Solid State Comm.*, **21**, 679–681

Fleetwood, D. M. and Giordano, N. (1983). 'Resistivity dependence of 1/f noise in metal films', *Phys. Rev. B*, **27**, 667–671

Flinn, I. (1968). 'Extent of the 1/f noise spectrum', *Nature*, **219**, 1356–1357

Frederick, H. A. (1931). 'The development of the microphone', *J. Acoust. Soc. Am.*, **3**, Supplement to No. 1 (July)

Hanafi, H. I. and van der Ziel, A. (1978). 'Flicker noise due to grain boundaries in n-type $Hg_{1-x}Cd_xTe$', *Solid State Electronics*, **21**, 1019–1021

Handel, P. H. (1975). '1/f noise—an infrared phenomenon', *Phys. Rev. Lett.*, **34**, 1492–1495

Hooge, F. N. (1969). '1/f noise is no surface effect', *Phys. Lett. A*, **29**, 139–140

Hooge, F. N. and Gaal, J. L. M. (1971). 'Fluctuations with a 1/f spectrum in the conductance of ionic solutions and in the voltage of concentration cells', *Philips Res. Rep.*, **26**, 77–90

Hooge, F. N. and Hoppenbrouwers, A. M. H. (1969). '1/f noise in continuous thin gold films', *Physica*, **45**, 386–392

Hooge, F. N., Kedzia, J. and Vandamme, L. K. J. (1979). 'Boundary scattering and 1/f noise', *J. Appl. Phys.*, **50**, 8087–8089

Hoppenbrouwers, A. M. H. and Hooge, F. N. (1970). '1/f noise of spreading resistances', *Philips Res. Rep.*, **25**, 69–80

Jones, B. K. and Francis, J. D. (1975). 'Direct correlation between 1/f and other noise sources', *J. Phys. D*, **8**, 1172–1176

Kedzia, J. and Vandamme, L. K. J. (1978). '1/f noise in liquid and solid gallium', *Phys. Lett.*, **66A**, 313–314

Ketchen, M. B. and Clarke, J. (1978). 'Temperature fluctuations in freely suspended tin films at the superconducting transition', *Phys. Rev. B*, **17**, 114–121

Kim, S. K. and van der Ziel, A. (1980). 'Noise in hydrogenated amorphous silicon resistors', *Physica B + C*, **98 B + C**, 303–305

Kirby, P. L. (1965). 'The non-linearity of fixed resistors', *Electron. Engg.*, **37**, 722–726

Kleinpenning, T. G. M. (1974). '1/f noise in thermo e.m.f. of intrinsic and extrinsic semiconductors', *Physica*, **77**, 78–98

Kleinpenning, T. G. M. (1977). 'Theory of noise investigations on conductors with the four-probe method', *J. Appl. Phys.*, **48**, 2946–2949

Kogan, Sh. M. and Nagaev, K. E. (1982). 'Low frequency current noise and internal friction in solids', *Sov. Phys. Solid State (USA)*, **24**, 1921–1925

Macfarlane, G. G. (1950). 'A theory of contact noise in semiconductors', *Proc. Phys. Soc. B*, **63**, 807–814

Machlup, S. (1981). 'Earthquakes, thunderstorms and other 1/f noises', *Proceedings of 6th International Conference on Noise in Physical Systems* (Ed. P. H. E. Meijer, R. D. Mountain and R. J. Soulen, Jr), NBS Special Publication 614, pp. 157–160

McWhorter, A. L. (1957). '1/f noise and germanium surface properties', *Semiconductor Surface Physics* (Ed. R. H. Kingston), pp. 207–228

Mandelbrot, B. B. (1977). *Fractals: Form, Chance and Dimension*, Freeman; San Francisco

Mandelbrot, B. B. and van Ness, J. W. (1968). 'Fractional Brownian motions, fractional noises and applications', *SIAM Rev.*, **10**, 422–437

Marinari, E., Parisi, G., Ruelle, D. and Windey, P. (1983). 'Random walk in a random environment and 1/f noise', *Phys. Rev. Lett.*, **50**, 1223–1225

Montgomery, H. C. (1949). Reported in an anonymous paper, 'Semiconductor Rectifiers', *Elect. Engg.*, **68**, 865–872

Musha, T. (1981). '1/f fluctuations in biological systems', *Proceedings of 6th International Conference on Noise in Physical Systems* (Ed. P. H. E. Meijer, R. D. Mountain and R. J. Soulen, Jr), NBS Special Publication 614, pp. 143–146

Musha, T., Sugita, K. and Kaneko, M. (1983). '1/f noise in aqueous $CuSO_4$ solution', *Noise in Physical Systems and 1/f Noise* (Proceedings of 7th International Conference on Noise in Physical Systems and 3rd International Conference on 1/f Noise. Ed. M. Savelli, G. Lecoy and J-P. Nougier), North-Holland; Amsterdam, pp. 389–392

Nelkin, N. and Tremblay, A-M. S. (1981). 'Deviation of 1/f voltage fluctuations from scale-similar Gaussian behavior', *J. Stat. Phys.*, **25**, 253–268

Neudeck, G. W. and Kriegel, M. H. (1978). 'Noise measurements in electron-beam-evaporated amorphous silicon thin films', *Thin Solid Films*, **53**, 209–215

Ngai, K. L., Jonsher, A. K. and White, C. T. (1979). 'On the origin of the universal dielectric response in condensed matter', *Nature*, **277**, 185–189

Otto, R. (1935). 'The noise of carbon microphones' (in German), *Hochfreq. Tech. u. Elektroakus.*, **45**, 187–198

Press, W. H. (1978). 'Flicker noises in astronomy and elsewhere', *Comments Astrophys.*, **7**, 103–119

Radeka, V. (1969). '1/|f| noise in physical measurements', *IEEE Trans.*, **NS-16**, 17–35

Richardson, J. M. (1950). 'The linear theory of fluctuations arising from diffusional mechanisms—an attempt at a theory of contact noise', *Bell Syst. Tech. J.*, **29**, 117–141

Sayer, M. and Prasad, E. (1979). 'Electrical noise in semiconducting oxide glasses', *J. Non-cryst. Solids*, **33**, 345–349

Schönwald, B. (1932). 'Electrical and optical properties of semiconductors' (in German), *Ann. Phys.*, **15**, 395–421

Scofield, J. H. and Webb, W. W. (1983). 'Observations of low frequency current noise in niobium films—electro diffusion noise and the absence of 1/f noise', *Noise in Physical Systems and 1/f Noise* (Proceedings of 7th International Conference on Noise in Physical Systems and 3rd International Conference on 1/f Noise. Ed. M. Savelli, G. Lecoy and J-P. Nougier), North-Holland; Amsterdam, pp. 405–408

Sinai, Ia. G. (1982). 'Lorentz gas and random walks', *Proceedings of Berlin Conference on Mathematical Problems in Theoretical Physics* (Ed. R. Schrader, R. Seiler and D. A. Uhlenbrock), Springer-Verlag; Berlin, pp. 12–14

Stoisiek, M. and Wolff, D. (1976). 'Recent investigations on the stationarity of 1/f noise', *J. Appl. Phys.*, **47**, 362–364

Stroeken, J. T. M. and Kleinpenning, T. G. M. (1976). '1/f noise of deformed crystals', *J. Appl. Phys.*, **47**, 4691–4692

Suh, C. H., van der Ziel, A. and Jindal, R. P. (1981). '1/f noise in GaAs MESFETS', *Proceedings of 6th International Conference on Noise in Physical Systems* (Ed. P. H. E. Meijer, R. D. Mountain and R. J. Soulen, Jr), NBS Special Publication 614, pp. 236–239

Tandon, J. L. and Bilger, H. R. (1976). '1/f noise as a non-stationary process: experimental evidence and some analytical conditions', *J. Appl. Phys.*, **47**, 1697–1701

Tolman, R. C. (1938). *The Principles of Statistical Mechanics*, Oxford University Press; London

Van Calster, A., van den Eede, L., de Molder, S. and de Kayser, A. (1983). '1/f noise in cermet and metanet resistors', *Noise in Physical Systems and 1/f Noise* (Proceedings of 7th International Conference on Noise in Physical Systems and 3rd International Conference on 1/f Noise. Ed. M. Savelli, G. Lecoy and J-P. Nougier), North-Holland; Amsterdam, pp. 193–195

Vandamme, L. K. J. and Van Ruyven, L. J. (1983). '1/f noise used as a reliability test for diode lasers', *Noise in Physical Systems and 1/f Noise* (Proceedings of 7th International Conference on Noise in Physical Systems and 3rd International Conference on 1/f Noise. Ed. M. Savelli, G. Lecoy and J-P. Nougier), North-Holland; Amsterdam, pp. 245–247

Van den Berg, R. J., de Vos, A. and de Goede, J. (1981). 'Concentration fluctuations in small volumes of ionic solutions', *Proceedings of 6th International Conference on Noise in Physical Systems* (Ed. P. H. E. Meijer, R. D. Mountain and R. J. Soulen, Jr), NBS Special Publication 614, pp. 217–220

Van der Ziel, A. (1950). 'On the noise spectra of semiconductor noise and of flicker effect', *Physica*, **16**, 359–372

Van de Voorde, P., Iddings, C. K., Love, F. W. and Halford, D. (1979). 'Structure in the flicker noise power spectrum of n-InSb', *Phys. Rev. B*, **19**, 4121–4124

Van Vliet, C. M. and Handel, P. (1982). 'A new transform theorem for stochastic processes with special application to counting statistics', *Physica*, **113A**, 261–276

Van Vliet, K. M., Handel, P. and van der Ziel, A. (1981). 'Superstatistical emission noise', *Physica*, **108A**, 511–530

Voss, R. F. (1978). 'Linearity of 1/f noise mechanisms', *Phys. Rev. Lett.*, **40**, 913–916

Voss, R. F. and Clarke, J. (1976). 'Flicker 1/f noise: equilibrium temperature and resistance fluctuations', *Phys. Rev. B*, **13**, 556–573

Voss, R. F. and Clarke, J. (1978). '"1/f noise" in music: music from 1/f noise', *J. Acoust. Soc. Am.*, **63**, 258–263

Weissman, M. B., Black, R. D. and Snow, W. M. (1982). 'Calculation of experimental implications of tensor properties of resistance fluctuations', *J. Appl. Phys.*, **53**, 6276–6279

Weissman, M. B., Black, R. D. and Restle, P. J. (1983). '1/f noise in silicon-on-sapphire', *Noise in Physical Systems and 1/f Noise* (Proceedings of 7th International Conference on Noise in Physical Systems and 3rd International Conference on 1/f Noise. Ed. M. Savelli, G. Lecoy and J-P. Nougier), North-Holland; Amsterdam, pp. 197–200

Chapter 3

Noise in Ferromagnetic and Ferroelectric Materials

3.1 Introduction

These two types of material are taken together because they both exhibit (1) spontaneous polarisation with a domain structure and (2) under the application of an external field (magnetic or electric respectively) discontinuous changes in polarisation which give rise to pulses known as Barkhausen noise in an associated electrical circuit. In the case of ferromagnetic materials, the specimen must be surrounded by a coil in which will appear a voltage proportional to dB/dt; and a ferroelectric specimen is used as the dielectric of a capacitor, the current to which will be proportional to dD/dt. It appears, however, that the mechanism of change of polarisation by re-arrangement of domains is different in the two cases.

3.2 Ferromagnetic materials: Barkhausen noise and domains

Bittel (1969) pointed out that three types of noise are associated with ferromagnetic material: (1) With magnetisation and temperature constant, there must be Johnson noise associated with the loss; (2) with constant magnetisation but variable temperature there is a phenomenon known as 'excess noise'; and (3) variable magnetisation (at assumed constant temperature) causes Barkhausen noise.

The last was first reported by Barkhausen (1919) and is now known to be related to domain structure, a ferromagnetic material being one which is spontaneously magnetised (at temperatures below its Curie point), though in variously oriented domains which are so arranged as to show little external field. The arrangement and sizes of domains are such as to minimise the *total* potential energy which is made up of the following components:

(1) External field energy, which is dependent on the extent to which domains fail to form closed magnetic circuits. In the classical model for polycrystalline material in the unmagnetised state, there are

many small domains which appear to be randomly oriented. (The orientation cannot, in fact, be random, for there would then be an average resultant in any plane proportional to the square root of the number of components, according to Rayleigh's random-vector theorem.) In single-crystal specimens there may be a few large domains which are systematically arranged to produce closed magnetic circuits.

(2) Magnetostriction energy, which is minimised by having many small domains.

(3) Domain-wall energy, which in general is proportional to the total area of domain walls. (The energy per unit area of the wall, together with the thickness of the wall, are estimated by minimising the sum of the magnetic anisotropy energy and the magnetic interaction energy within the wall.)

It will be seen that (2) (and possibly (1)) favours small domains while (3) favours large domains, so that between them they define a domain size for minimum total energy. If these were the only factors, the changes in magnetisation under an applied field would be reversible. But Néel (1946) has pointed out that the presence of minute non-magnetic inclusions in a ferromagnetic material will produce free poles within the material, with corresponding magnetostatic energy. (Compare the use of precipitation-hardened alloys for permanent magnets.) In addition, mechanical strain can produce changes in *direction* of magnetisation and this also implies the presence of free poles, a mechanism which applies particularly to nickel alloys. The presence of these local defects in the structure contributes to the division of the crystal into smaller domains, because the defects act as nuclei for the setting up of new directions of magnetisation. At one time it was thought that the observed coherence of Barkhausen noise along the length of a specimen implied domain lengths of a centimetre or so; but the coherence of magnetic-field impulses over such lengths is due to eddy currents. It appears from powder patterns that the linear dimensions of domains are usually of the order of 0.01 to 0.1 cm, whereas Mazetti and Montalenti (1962), in discussing Barkhausen noise, suggested a domain volume of 10^{-9} cm^3. There are three points to remember when comparing these figures:

(1) recorded powder patterns tend to be from larger than average domains;

(2) the domains in powder patterns do not approximate to cubes but are nearer to needles (so that the volume is less than the cube of the greatest dimension);

(3) Barkhausen noise is usually due to *changes* of domain volume, so that it does not give a direct indication of volume.

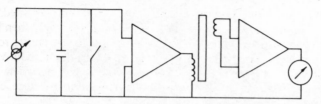

Fig. 3.1 Schematic circuit for measurement of Barkhausen noise

A general review of domain theory was given by Kittel and Galt (1956), but there has been a great deal of later work on the movement of domain walls. Hubert (1974) has reviewed in detail the structure of domain walls in various media, including ferromagnetics, superconductors, antiferromagnetics, diamagnetics, ferroelectrics and liquid crystals.

The 'magnetisation' of a specimen of ferromagnetic material is brought about by re-arranging the domains and proceeds in three stages. Initially the boundaries between domains move reversibly in such a way as to enlarge those domains which are pointing at less than 90° from the direction of the applied field, at the expense of those pointing in other directions. This occurs in the region A in Fig. 3.2, where Rayleigh's law applies, the permeability following the law

$$\mu = \alpha + \beta H \tag{3.1}$$

This also applies to a small change of field about any given point on the magnetisation curve, i.e. to a small minor loop. In the second stage, domain boundaries jump discontinuously and irreversibly past the Néel defects by which they were initially held (Fig. 3.4). These jumps constitute the Barkhausen noise and occur mainly on the steep part of the curve, region B in Fig. 3.2. In the third stage, region C, the direction of magnetisation of the domains is pulled away from the preferred directions defined by crystal axes and to the direction of the applied field. At first it was questioned whether the Barkhausen jumps were due to irreversible movements of domain walls or to rotation of complete domains. Several experimenters, following Williams and Shockley (1949), have used single-crystal material in 'picture frame' form, as shown in Fig. 3.3, with a minimum number of domains. They found the usual Barkhausen noise during magnetisation, which must have been due to irreversible movements of 180° domain walls. Tebble and Newhouse (1953) also used single-crystal material, though not in picture-frame form, and concluded decisively that the Barkhausen noise in single-crystal specimens was due to movements of 180° walls.

In polycrystalline material, however, some of the domains must

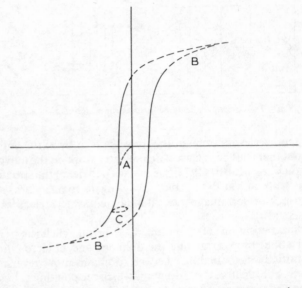

Fig. 3.2 Barkhausen noise is strongest in the part of the magnetisation curve shown in full line. It occurs with much smaller magnitude in the regions A, B and C shown dotted

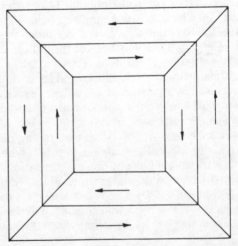

Fig. 3.3 Domains in 'picture frame' single-crystal specimen

initially be nearer 90° than 180° from the direction of the applied field. But magnetisation can occur through nucleation and expansion of domains in the field direction, so that Barkhausen noise will still be due to movements of domain walls. Deimel *et al.* (1977) used polycrystalline wires of very pure iron (better than 99.9%) of 0.5 mm diameter in an investigation of the effect of grain size on Barkhausen noise. They concluded that at room temperature the contributions from 90° domains were dominant, but at 77 K, where the total Barkhausen noise was less, the contributions from 90° and 180° domains were comparable. The volumes which change their magnetisation suddenly in the Barkhausen jumps are therefore usually *increments* of domain volume, rather than complete domains. The size of jump in the magnetisation of the specimen, given the saturation magnetisation of the material, leads to an estimate of the volume involved. This is usually estimated to be between 10^{-10} and 10^{-8} cm^3. Tebble (1955) deduced that non-magnetic inclusions will cause Barkhausen jumps only if their linear dimensions are of the same order as the thickness of a domain wall, say between 1 and 2×10^{-5} cm in iron. Smaller inclusions will not be effective in restraining wall motion and larger inclusions will be surrounded by subsidiary domain structures so that the outer domain walls move reversibly around them. In nickel and nickel alloys, however, the defects are not usually non-magnetic inclusions but free poles where the direction of magnetisation changes as a result of mechanical strains of the crystal structure; and instead of size of inclusion one must consider the energy associated with the free poles at the defects. Barkhausen jumps will occur whenever the local rate of change of wall energy as the wall moves is greater than the rate of change of energy associated with the overall magnetic moment.

The third stage, the rotation of domains towards the exact direction of the applied field, is mainly reversible and therefore not a source of Barkhausen noise. In square-loop materials, e.g. as used in transductors, this third stage is practically eliminated by orienting the crystallites (in polycrystalline material) by cold rolling so that one of the easy directions of magnetisation always lies along the axis of the applied field. But such square-loop materials have large hysteresis loss and are not usually employed where noise level is critical (except in magnetic modulators as used in fluxgate magnetometers).

Three questions have to be answered: (1) Does the sum of Barkhausen pulses correspond with the total of the irreversible change? (2) Is the mechanism deterministic on the microscopic scale, so that if the material is taken round a repeated cycle of applied field the Barkhausen noise can be Fourier analysed into harmonics of the repetition frequency? (3) How does the intensity of Barkhausen noise

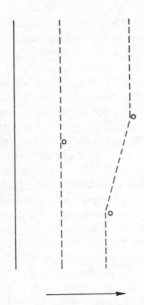

Fig. 3.4 Pinning of a domain wall by crystal defects. The dotted lines represent successive positions of a wall moving in the direction of the arrow

vary with the magnetic and physical characteristics of the material, such as permeability, coercive force and grain size?

A schematic arrangement for observing Barkhausen noise is shown in Fig. 3.1, the rate of change of magnetising current being controlled by the rate of change of voltage on the capacitor, which in turn controls the current output of the amplifier. One would in practice use an adaptation of an electronically controlled power supply to supply an adjustable constant current to the capacitor. If individual Barkhausen pulses are to be detected, the magnetising field must be varied very slowly: Bozorth (1929), who used a temperature-limited thermionic diode as a source of constant current which could be varied by varying the cathode temperature, quotes rates between 0.001 and 10 oersted per second. In order to obtain a true measure of the effect with 'd.c.' magnetisation, Bozorth observed the number of Barkhausen pulses at different rates of change of the field and extrapolated to zero rate of change. Two opposed detector coils may be used, so as to balance out any effect of fluctuations in the magnetising current. These two coils should be sufficiently far apart on the specimen under investigation that the Barkhausen pulses in them may be uncorrelated; and the two sets of pulses may then be combined, or since the pulses are unidirectional the pulses from one of the two opposed coils can be suppressed by a rectifier, after amplification. (This would require the amplifier to have a d.c.

response, at least up to the stage at which this rectifier is applied. It is only if the pulses were very widely separated that one could neglect the error due to a.c. coupling.)

The use of two opposed coils to balance out interference may be legitimate in slow, 'd.c.', magnetisation, but it is not recommended if the magnetic material is cycled at any frequency from a few Hz to a few kHz. The non-linearity of the magnetic material will produce harmonics of the exciting frequency and one cannot rely on the balance between the two coils being good enough to reduce these harmonics below the Barkhausen noise at the same frequency.

A minor phenomenon is the *transverse* Barkhausen effect (Bozorth and Dillinger, 1932). If in the demagnetised state the domains are oriented in various directions, then change towards the field direction, either by movement of domain walls or change of orientation of domains, will reduce the magnetic moment transverse to the field and the resulting Barkhausen noise is mainly less than 10% of the axial noise.

3.3 The spectrum of Barkhausen noise

Recent mathematical work has been concerned with providing a quantitative theoretical account of Barkhausen phenomena and noise spectra in terms of movements and velocities of domain walls[*]. The speed of movement of a domain wall in a metallic body is controlled by local eddy currents, while more widespread eddy currents control the shape (in both time and space) of the pulse which may be detected outside the specimen. (The spatial extension by eddy currents leads to a spurious correlation in an axial direction which misled some early experimenters into thinking that the domain extended over the correlation distance.)

Tebble *et al.* (1950) have calculated the shape of the pulse of e.m.f. resulting from a sudden change of magnetic moment in a small volume within a cylindrical conductor. The change of magnetic moment itself is believed to occur in a very short time; and the pulse of e.m.f. in a search coil surrounding the cylinder is controlled by eddy currents and is then given by

$$e = -\frac{d\varphi}{dt} = \frac{\varphi_0}{2\pi\mu\sigma a^2} \sum_{n=1}^{\infty} \frac{\lambda_n J_0(\lambda_n b/a)}{J_0(\lambda_n)} \exp(-\lambda_n^2 \xi) \tag{3.2}$$

where μ, σ and a are the permeability, conductivity and radius of the

---

[*] Russian work of this kind is reported in the Proceedings of the All-Union Seminar on the Barkhausen Effect which is available in the American translation, *Bulletin of the Academy of Sciences of the USSR, Physical Series*, Vol. 45, No. 9.

specimen; b is the radius at which the change of magnetic moment is situated; λ_n is the nth root of $J_n(\lambda) = 0$; and ξ is a scaled time equal to $t/4\pi\mu a^2\sigma$. It follows that the e.m.f. is not a simple exponential pulse, but is represented by a sum of exponentials. An effective time-constant τ_0 for a change situated on the axis can be defined by relating the maximum e.m.f. to the flux change, and evaluated thus:

$$e_{max} = \varphi_0/\tau_0$$
$$\tau_0 = 1.04\pi\mu a^2\sigma \tag{3.3}$$

The observed distribution of pulse widths over a small change in H near the value equal to the coercivity was consistent with the relationship $\tau \propto a^2$. In addition, the external pulse may be reduced by the demagnetising effect of a finite specimen, depending on its length/diameter ratio (Tebble and Newhouse, 1953).

The first thought was that the sum of ΔB corresponding to Barkhausen voltage pulses should be equal to the total change of flux in a particular change of magnetisation. But experimentally this accounted for anything between 10% and 90% of the change of flux between magnetisation to saturation in opposite directions. This is attributed to clustering of pulses whereby the detecting apparatus may count a cluster as one broad pulse and so under-estimate the sum of pulse heights. (These clustered pulses are not to be confused with the 'giant pulses' which are sometimes produced by the complete reversal of a sizable domain instead of the limited movement of a domain wall.) But, apart from voltage pulses, each change ΔB in flux is associated with a change ΔH in magnetising field; and the sum round a cycle of magnetisation of the energy pulses $\Delta B \, \Delta H$ should be equal to the hysteresis loss round the cycle.

This leads to the concept of an energy spectrum (in contrast to the usual power spectrum) and the possibility of specifying the Barkhausen noise of a material in terms of noise energy per cycle of magnetisation. Since squared voltage is now to be integrated with respect to time, whereas voltage only was integrated to predict flux change, observations are not invalidated by the occurrence of clustering of pulses. The relations are the if $U(t)$ is the instantaneous noise voltage (embracing all frequencies) and $W(f)$ is the power spectrum,

$$\overline{U^2}(t) = \int W(f)\,\mathrm{d}f \tag{3.4}$$

but if $E(f)$ is the energy spectrum referred to one cycle of magnetisation

$$\int_0^\tau U^2(t)\,\mathrm{d}t = \int E(f)\,\mathrm{d}f \tag{3.5}$$

The noise characteristic of a material can then be specified by the value of $E(f)/n^2A$ which is in units of $\mathrm{V}^2\ \mathrm{s}^2\ \mathrm{turn}^{-2}\ \mathrm{cm}^{-2}$ and refers to 1-Hz bandwidth, 1 turn of detector coil and unit area of cross-section of specimen. Figures for various materials were collected by Mazetti and Montalenti (1963) and were mostly of the order of 10^{-13} at a magnetisation cycle frequency of 0.2 Hz, though no Barkhausen noise could be detected in mumetal having $\mu_{max} = 300{,}000$. Bittel (1969) calculated that the independent reversal of a single spin would give a figure of 2×10^{-30} in these units, which gives a ratio of the order of 10^{17} between this and the observed noise. If the atoms are equally spaced on all three axes the observed jump would appear to involve, for example, a cubic array with about 0.5×10^6 atoms in each direction. This corresponds to coherence over a distance of about 2×10^{-2} cm, which is comparable with the supposed dimension of a domain. Since it is now supposed that Barkhausen noise is due to the limited motion of domain walls, rather than the rotation of complete domains, this implies some clustering of pulses, i.e. the observed jump is due to the correlated occurrences of more than one basic event, since the basic events are smaller than would correspond with the observed pulse.

In Equations (3.4) and (3.5) the integral with respect to frequency has been left open. In principle, it should run from zero to infinity but, in practice, the spectrum of Barkhausen noise does not extend appreciably above 50 kHz. For total magnitude, Mazetti and Montalenti (1963) quote as typical an e.m.f. of 50 μV in a detector coil of 1,000 turns on a toroidal specimen having its magnetisation cycled at 0.1 Hz. The noise power (squared voltage) is approximately proportional to the frequency at which the magnetisation is cycled: this follows from the fact that each cycle of magnetisation provides its quota of energy, and the proportionality can be seen in Fig. 3.5. For frequencies more than two or three times the frequency of the magnetising current and amplitude at least sufficient to reach the knee of the magnetisation characteristic, the power spectra may be represented approximately by

$$\varphi_p(f) = \frac{N_t^2}{\pi}\,\frac{Kf_m}{a^2 + (2\pi f)^2}\left[1 - \frac{1}{2 + 2 \times 10^{-5}(f/f_m)^2}\right] \tag{3.6}$$

where K and a are constants of the specimen, N_t is the number of turns of the detector winding and f_m is the frequency of the magnetising current. K is a material constant but a is closely related to the reluctance of the specimen, so that the low-frequency turnover of the

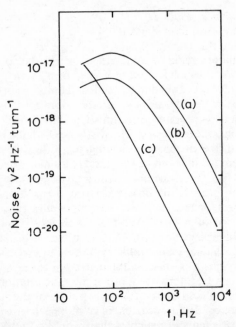

Fig. 3.5 Frequency spectra of Barkhausen noise. Curves (a) and (b) are for a cylindrical specimen, 3.5 cm long and 0.5 mm diameter, with $f_m = 0.2$ and 0.005 Hz respectively. Curve (c) is for a toroid with $f_m = 0.1$ Hz. Both specimens were of pure iron, annealed in hydrogen. (After Mazetti and Montalenti)

noise curves for a cylindrical specimen is largely eliminated by the use of a toroidal specimen. (The low-frequency turnover is due to a limitation of the changes in magnetisation of the specimen by the demagnetising effect of free poles.) For frequencies below three times f_m and for smaller amplitudes the noise is reduced by a factor β: see Mazetti and Montalenti (1962), Equation (2) and Fig. 2, following the work of Bunkin (1956).

Spectra of very similar shape were found by Wiegman (1977) for thin (metallic) magnetic films, though here the duration of a Barkhausen pulse was fixed by domain wall mobility rather than by gross eddy currents. The film thicknesses used were 400 to 2400 Angstrom units, which meant that the domain pattern was two-dimensional; and the type of domain wall changed from Bloch at the greater thicknesses to cross-tie and Néel as the thickness diminished below about 900 Å. The spectra could be represented approximately by

$$E(f) = \frac{A}{1 + (f/f_c)^n}$$

where n was always about 1.7 and f_c ranged between 30 and 400 Hz for film thickness less than 900 Å but was up to 5×10^3 Hz for thicker films. Note that this corresponds to a domain wall movement much slower than that inferred for bulk material.

3.4 Thermal effects

If all the changes in arrangement of domains were strictly deterministic, the flux change from repeated cycling of magnetisation could be represented by Fourier analysis into a line spectrum of harmonics of f_m and there would be no noise. This appears to have been first noted by Grachev (1950), who showed experimentally that, although harmonics of f_m could be detected, there was also a continuous spectrum of noise. (Most experimenters used so low a value of f_m that its harmonics would not be detectable in the frequency band used for observation of noise.) This is a direct indication that the movement of domain walls is influenced by random thermal agitation as well as by the applied magnetising field: a fluctuation in thermal energy may enable a domain wall to overcome a potential barrier although the externally applied field does not supply sufficient energy. Bukharov and Koluchevskii (1981) considered that fluctuations occurred in both the state of magnetisation and the coercivity of a ferromagnetic body, the two fluctuations being in quadrature; and they studied the correlation between them both theoretically and experimentally. They balanced out the component of flux change at f_m to a first order by using two toroidal cores, but since they were measuring the correlation between the in-phase and quadrature components of fluctuation the exact suppression of the output at their f_m of 1.4 kHz was not essential. They found what we would call $1/\Delta f$ noise (see Chapter 2) within a range of 5 Hz from f_m. A mathematical analysis by Fedosov and Chursima (1981) found that the spectrum of noise should be flat, very close to the harmonic of f_m, but $1/\Delta f$ for greater frequency differences. In most of the classical work on Barkhausen noise the frequency f_m (and therefore the separation between harmonics) was less than the bandwidth used in the analysis of noise, so that this $1/\Delta f$ effect would not have been detectable.

This thermal fluctuation also causes fluctuations of magnetisation in the absence of an applied field, as was noted by Brophy in 1958. Such fluctuation is to be expected as a form of Johnson noise associated with the loss in the magnetic material, expressed by Bittel (1969) in the form

$$W(f) = 4kT\mu''\omega L \qquad (3.7)$$

This is correct at any point in the magnetisation cycle provided one uses the appropriate value of μ'', the loss component of the complex permeability $\mu = \mu' + j\mu''$. This had been verified experimentally by Bittel and Lüttgemeier (1963) for both toroids wound with high-permeability tape and ferrite cores, though the resistance of windings accounted for 36% to 80% of the total loss resistance. In order to avoid complications from non-linearity they connected a capacitance C across the coil and compared the mean fluctuation energy stored in C with the value kT/C predicted by equipartition theory. Bittel identified the means of transfer of thermal energy as (1) spin waves and (2) conduction electrons. The magnetic after-effect—sometimes known as the Jordan effect—also depends on thermal agitation. After removal of a magnetic field from a ferromagnetic body, the latter may slowly relax with generation of Barkhausen noise. (A practical example which recognises this relaxation on a macroscopic scale is the 'stabilisation' of permanent magnets by applying a small reverse field after magnetisation to saturation.) The third manifestation of thermal fluctuations is in the *excess noise* which is found when the temperature is changed at constant field. Bittel assumed that geometrical change of structure was involved, since Invar (with no thermal change in dimensions) did not show excess noise.

Since the characteristics of the Barkhausen noise depend on the structure of the material, it is natural to use the noise as a diagnostic tool for the non-destructive testing of ferromagnetic objects. The study of domain walls as a means of probing relaxation phenomena in solids was discussed by Hubert (1974), in Section 12 of his work, and it is now a routine method of examining steel castings (Schroder and Saijja, 1982).

3.5 Amorphous solids

Normal (metallic) ferromagnetic materials are polycrystalline, but since it has been found possible to produce metals in amorphous form (by rapid chilling from the molten state) there has been interest in the magnetic properties of amorphous metals. Unfortunately it is necessary to use special alloys to make it practicable to prepare thin metal strip in the amorphous state: boron is one of the usual additives to iron and nickel alloys. Therefore the amorphous alloy is not directly comparable with any normal ferromagnetic alloy, because of its constitution as well as the amorphous state. Fiorillo *et al.* (1977) studied the Barkhausen noise in an alloy with the atomic percentage constitution $Fe_{40}Ni_{40}P_{14}B_6$. They found firstly that there was a clustering of Barkhausen pulses and secondly that the temperature dependence of Barkhausen noise was the same as that of static (hysteresis) loss. Since then, there has been much development of magnetic amorphous materials because they can have high permeability, low coercive force (therefore low hysteresis loss) and high electrical resistivity; and it is hoped eventually to use amorphous ferromagnetics for low-loss cores of large power transformers. The absence of a crystalline structure means that there is no anisotropy (unless due to mechanical strain) and little restraint on the movement of domain walls. It has been claimed that during cyclic magnetisation the only (or dominant?) restraint on their movement is provided by eddy currents; but Barkhausen noise has been observed during slow magnetisation and pinning of domain walls is believed to occur mainly where they intersect the surface of the material. In any case, the Barkhausen noise of amorphous ferromagnetic materials can be expected to be small, corresponding with their small hysteresis loss. The general review of amorphous metals by Chen (1980) includes a section on magnetic properties.

3.6 Ferrites and semiconductors

Ferrites form a distinct class, firstly because they have very low electrical conductivity and secondly because they are ferrimagnetic rather than ferromagnetic. Their hysteresis loss is large compared with that of the best 'soft' ferromagnetic alloys, so they are not usually employed at the fairly low frequencies at which Barkhausen noise can be expected. (Toroids wound with thin metal tape have been used for signals up to MHz frequencies.) However, a paper by Drokin et al. (1981) which was primarily concerned with the ordering of cobalt ions in nickel-cobalt and lithium-cobalt ferrites, reported Barkhausen jumps having a duration of 25 µs to 100 ms and occurring mainly on the steepest part of the magnetisation curve. This is similar to the behaviour of ferromagnetics, for which the Barkhausen noise occurs mainly below 50 kHz.

There are also magnetic semiconductors which are affected by light, as discussed by Lems et al. (1968), and for which the first experimental observation of the effect of light on Barkhausen noise is claimed by Veselago et al. (1981), who worked with light from a filament lamp passed through a monochromator to illuminate a 2-mm-diameter torus cut from a single crystal of $Cd_2Cr_2Se_4$. The magnetising current was at a frequency of 20 Hz and varied in magnitude to produce a field from zero to 0.30 Oe. The effect of light was most marked at fields around 0.05 Oe and was negligible above 0.3 Oe.

3.7 Applications of ferromagnetic materials

In the consideration of the effect of Barkhausen noise on signals it is convenient to distinguish three regimes: (1) the signal is so small that only the Nyquist noise is effective; (2) the signal is small enough to keep the swing in magnetisation within the Rayleigh region; and (3) the signal or an associated carrier is large enough for the magnetising cycle to include a region where major Barkhausen noise is generated.

Regime (1) does not require special discussion because the thermal noise is the same whether originating in magnetic losses or in some other loss mechanism in the circuit.

The practical question is whether small-signal operation of a component with a ferromagnetic core can be kept within the Rayleigh region of the latter. (This also covers a small variation about any fixed point on the magnetisation cycle, as might result from a d.c. bias of the core.) A reversible swing of 0.01 Oe in 1 cm^3 of material with a reversible permeability of 1000 would represent an energy change of approximately 4×10^{-10} J; and this is so far above the thermal equipartition value of 4×10^{-23} J in a system of unit bandwidth that any low-level signal (i.e. not a high-power signal for audio output or the like) should fall within this range where the Barkhausen noise is small.

In relation to regime (3), Warren (1961) analysed the effect of Barkhausen noise in transformers and concluded that the mean square* voltage per turn of winding was proportional to the area of cross-section of the core and inversely proportional to the length of magnetic circuit.

The possibility of keeping the flux change within the reversible range is the only consideration in transformers, but in filters it may be asked whether the circuit design has any effect. Little can in fact be done by way of design of filter circuit: the total inductive stored energy is practically fixed when the power, image impedance and frequency range are prescribed, as may be seen from the examples of simplified low-pass and high-pass filters, using the approximations of constant characteristic impedance, etc. In the former (Fig. 3.6) the power transmitted is $P = i^2 Z_0 = i^2 \sqrt{(L/C)}$. It follows that twice the inductive stored energy may be written

$$i^2 L = PL\sqrt{(C/L)} = P\sqrt{(LC)} = P/\omega_c \qquad (3.7)$$

In formula (3.7) the energy is correctly shown as equivalent to a power divided by a frequency; and the noise power, which depends on the frequency with which the energy flows in and out of the inductance,

---

* Warren actually expressed his results in terms of root-mean-square.

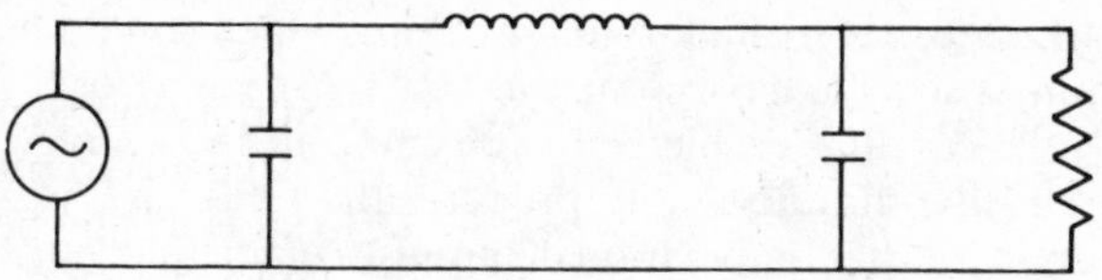

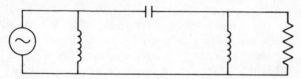

Fig. 3.6 Schematic low-pass filter

Fig. 3.7 Schematic high-pass filter

will be proportional to P and to the working frequency:

$$N \propto P(\omega/\omega_c) \qquad (3.8)$$

In the high-pass filter (Fig. 3.7) the current in the coil is $V/\omega L$ and the power transmitted is $P = V^2/Z_0 = V^2\sqrt{(C/L)}$. It follows that the stored energy may be written

$$i^2 L = (V^2/\omega^2 L^2)L = \frac{P\sqrt{(L/C)}}{\omega^2 L} = P(\omega_c/\omega^2)$$

As before, the noise is proportional to the working frequency, so that

$$N \propto P(\omega_c/\omega) \qquad \cdot (3.9)$$

The difference between low- and high-pass cases arises from the fact that all the load current must flow through the inductance in the former but not in the latter.

The obvious example of regime (3) is the magnetic modulator, in which the core is necessarily carried through the whole range of the Barkhausen effect during each cycle of the carrier frequency. The noise power output is proportional to the carrier frequency, but it is likely that the signal power output will also be proportional to carrier frequency so that the signal-to-noise ratio should be independent of the carrier frequency. For a complete swing between saturation in the two directions one always has $\Sigma \, \Delta B \simeq 2B_{max}$, since the contribution from reversible changes is usually a minor part. But if the individual steps occur at random time intervals the noise power is proportional to $\Sigma(\Delta B)^2$, and for fixed total excursion this can be minimized by making the steps numerous but of small size. The material will then be required to contain a large number of uniformly distributed defects, in contrast to material for small-signal working which should have few discontinuities so as to have comparatively large ranges of reversible change. If the core has a volume of 100 cm^3 and the average size of Barkhausen discontinuity (more properly the r.m.s. size, cf. Campbell's theorem for the combination of impulses of unequal magnitude, see Appendix) is 10^{-9} cm^3, the number of discontinuities should be 10^{11}. The Barkhausen noise should then be less than would be produced by a single abrupt change between positive and negative saturation by a factor of 10^{11}. Noble and Baxendall (1952) found a Barkhausen noise power of 10^{-17} W in an effective bandwidth of 1 Hz, but before making a comparison one must convert between energy and power in a narrow frequency band. Tebble et al. (1950) suggested that the actual change of moment occurs in a very short time (short compared with 5 μs), so that if one supposed the energy to be spread uniformly over a frequency band of 10^6 Hz one would arrive at a power ratio of $10^{11} \times 10^6 = 10^{17}$; and the exciting power for such a core would be of the order of 1 W. There are many assumptions in the above estimate, but at least it shows that it is possible for the observed value of Barkhausen noise in a magnetic modulator to be consistent with the theoretical studies.

Closely akin to the second-harmonic modulator/amplifier is the fluxgate magnetometer which is widely used for terrestrial geological surveys and for the exploration of magnetic fields in other parts of the solar system, from space vehicles. The fluxgate magnetometer resembles the second-harmonic magnetic amplifier in that its operation depends on the difference in onset of saturation in two pieces of magnetisable material which are driven in opposition by windings fed with a.c. and in parallel by the field to be measured. The difference is that in the amplifier the input field is generated by a signal current in an appropriate winding and the topology of ferromagnetic material and windings can be such as to contain all the fields within the cores. Methods of construction of fluxgate magnetometers, and their performance, are outlined in a review paper by Primdahl (1979); and the field to be measured may be either parallel to the a.c. exciting field or orthogonal to it. Owing to the driving of the magnetic material cyclically to saturation, the Barkhausen noise is dominant over the other two components which are (1) Nyquist noise corresponding to the resistance of the sensing winding plus the equivalent loss resistance of the iron and (2) excess noise (particularly during temperature change) due to domain relaxation (p. 72). The drive frequencies are commonly between 5 kHz and 15 kHz and the pass band in the output of the detector is from d.c. or a fraction of a hertz to a few hertz. Measured r.m.s. noise amplitudes in a bandwidth of 0.01 to 10 Hz range from 0.3 nT to 7.9 pT (1 pT $= 10^{-8}$ G). In addition to noise there may be an offset, but offset stability over periods from 24 hours to 6 months is usually better than 1 nT (10^{-5} G).

3.8 Ferroelectric materials

The existence of material capable of retaining permanent electric polarisation, directly analogous to the permanent magnetisation of ferromagnetic material, was recognised and discussed by Heaviside (1892). He suggested the term 'electrisation' for the production of permanent electric polarisation (as distinct from 'electrification' for the addition of charge) and 'electret' for the body analogous to a permanent magnet. The electrodes of an electret should be short-circuited when not in use, just as one uses a 'keeper' with a permanent magnet. Heaviside said that 'certain crystals, if no other bodies, are natural electrets'; and he noted the relevance of piezoelectric activity and that a pyroelectric crystal is a natural electret.

Eguchi (1925) and Gemant (1935) were more interested in the possibilities of synthetic electrets which might be of practical use as analogues of permanent magnets. They used various organic substances mixed with a hard wax which was cooled from the molten state in an electric field. Gemant described a typical specimen as 2.5 cm in diameter, 1.6 mm thick and cooled from the molten state in a field of 11 kV/cm. He found that all compositions showed immediately a surface charge opposite to that on the adjacent activating electrode, which he attributed to movement of ions through the liquid wax (Gemant, 1933). This surface charge would decay after a few days, but with some of the admixed substances there would remain a charge of the *opposite sign* which was a volume (polarisation) effect: it could not be changed by removing the surface of the electret. He attributed this (reversed) polarisation to a piezoelectric effect from the mechanical stresses due to the cooling of the wax. Thus Gemant, like Heaviside, recognised the relation between the piezoelectric effect and electrets.

The molecular mechanism of the ferroelectric effect is quite different from the atomic mechanism of ferromagnetism; and this is evidenced by the close relationship between ferroelectric, piezoelectric and pyroelectric effects, whereas there is no significant piezomagnetic effect. (Magnetostriction, like electrostriction, is a *quadratic* effect, the change in dimension being independent of the direction of the applied field. As it is a quadratic effect it is very small for small applied fields and there is no directly inverse effect.) Ferromagnetic moments result from unpaired electron spins; and the Curie temperature is that at which the disordering effect of thermal energy overcomes the ordering effect of quantum-mechanical exchange forces which produce ferromagnetic domains. The ferroelectric Curie point, on the other hand, always marks the transition from a higher-temperature symmetrical crystal structure, having no polar moment, to a lower-temperature asymmetric structure which gives each molecule a dipole moment. Owing to their asymmetry, ferroelectric materials show also piezoelectric, pyroelectric and electro-optic effects. Note that there may be further structural changes at lower temperatures, giving rise to anomalies in the dielectric constant, etc., at these other critical temperatures.

Since ferroelectric effects show as changes in electric displacement D, in contrast to changes in magnetic induction B for ferromagnetism, the experimental specimens are of a different shape. Whereas specimens for the study of magnetic Barkhausen noise may be either in the form of thin rods or ideally in the form of a closed circuit (either toroid or 'picture frame'), ferroelectric specimens are in the form of thin sheets. A typical specimen (Chynoweth, 1959) had a thickness of 10^{-1} mm and was about 3 mm square. Electrodes could be evaporated metal film (silver, gold or platinum—Chynoweth used platinum) or conducting liquids such as LiCl, to minimise the ohmic resistance between electrode and crystal.

There is also a complete difference between ferromagnetics and ferroelectrics in the mechanisms of domain formation and growth*. The boundary conditions at the 'pole' surface of an element of magnetic moment, such as a domain, merely require that there shall be an external magnetic field proportional to the discontinuity in magnetic polarisation; but a discontinuity in electric polarisation requires the presence of a *surface charge*, which is impossible within a perfect insulator. (Some anomalies in the behaviour of ferroelectrics have been attributed to the gradual accumulation of space charge within the body of the material.) Domains must therefore extend right through a specimen from one electrode to the other. (One of the usual substances for examination is barium titanate, in which domains can be observed optically.) The first stage in the process of polarisation of a depolarised ferroelectric is the formation of a cylindrical domain about 1 μm in diameter between the electrodes; and this is accompanied by a pulse of the order of 2×10^{-14} C with a rise time of a few microseconds. This domain usually remains static for several minutes; but the commencement of growth, if it occurs, is accompanied by a further pulse, of magnitude about 4×10^{-14} C and rise time around 100 μs. A similar pulse occurs when two domains coalesce, its magnitude being proportional to the area of wall which is eliminated. Miller (1960) suggested that when the two walls have

---

* For a general account of ferroelectrics, see Lines and Glass (1977).

come very close, the remaining thin volume between them which is polarised in the opposite sense will suddenly reverse on incorporation in the new enlarged domain. Miller observed pulses due to domain coalescence which were up to 2×10^{-12} C in magnitude. If the field was removed when the specimen was partially reversed, pulses continued to occur for some time, perhaps hours (cf. the magnetic after-effect). No pulses have been detected during the motion of a smooth domain wall: there appears to be nothing analogous to the pinning of magnetic domain walls by crystal defects. But there is some evidence that jerky motion of domain walls may be produced by irregularities in an evaporated metal electrode: the use of liquid electrodes avoids this possibility. Also, nucleation centres can sometimes be identified with visible crystal defects. An unexplained phenomenon is the creation by an applied field of a few domains of reversed polarity, accompanied by negative pulses of charge. These reversed domains sometimes occur reproducibly at sites which can be identified with either crystal or electrode defects. The maximum charge displacement on polarisation depends on the dielectric constant, which is a function of frequency, and the applied field, which is usually a few volts over a thickness of 0.1 mm. The total charge transferred in a typical specimen may then be about 10^{-8} C.

3.9 Pyroelectric effects

Ferroelectric materials in general are also pyroelectric, because their state of polarisation varies with temperature. The observable phenomenon is change of surface charge, corresponding to change of polarisation, with change of temperature. An output signal proportional to rate of change means that the responsivity of a pyroelectric device is basically independent of the frequency of modulation of the input: a higher signal frequency causes less change in temperature for the same amplitude of radiation change, but this is compensated by the higher rate of change with the higher frequency.

The natural application of pyroelectric materials is to thermal imaging, using radiation of too long wavelength to be detected by any form of photoelectric effect (liberation of electrons or photoconduction). Both TGS and ceramic material have been used for pyroelectric imaging (Blackburn et al., 1975; Higham and Wilkinson, 1975). Single-element TGS sensors had been made with a detectivity $D^* = 6 \times 10^8$ cm Hz$^{-1/2}$ W^{-1} at 10 Hz and 32 element arrays of TGS sensors for the detection of room temperature radiation had $D^*(300 \text{ K}, 10, 1)$ of the order of 2×10^7 cm Hz$^{-1/2}$ W^{-1}. But TGS is hygroscopic and, moreover, cannot in practice be used at temperatures above 35°C. It is also easily depolarised. The more robust

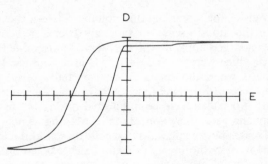

Fig. 3.8 Hysteresis loop for LATGS

ceramic detector could be made with $D^* = 1.2 \times 10^7$ cm Hz$^{-1/2}$ W^{-1}.

Interest in ferroelectrics was enhanced by the practical application of the pyroelectric property of triglycine sulphate (TGS) to radiation detection, particularly in the infra-red. An important feature of TGS is that by introducing a small proportion of laevo-alanine (10% in the solution from which crystals are grown results in the incorporation of 0.1% in the crystals, according to Keve *et al.*, 1971) the resulting crystals of LATGS are spontaneously polarised in one direction. This is because the TGS molecule has a plane of mirror symmetry, which allows the polar moment to point in either of two directions so that the bulk is neutral, but alanine lacks this symmetry, though its structure is otherwise similar to that of TGS; and a small proportion of alanine suffices to cause all the TGS to set in the direction established by the additive. The effect is obvious when one observes the hysteresis loop, that of the LATGS with alanine (Fig. 3.8) being shifted along the axis of E (which corresponds with the axis of H in a magnetic B/H loop). Note that with zero applied field E the polarisation P is nearly complete, so that for small applied signals there will be little noise of the Barkhausen type.

Many substances, including polymers, are ferroelectric to a greater or less degree and therefore are also piezoelectric and pyroelectric. The polymer PVF (poly-vinylidene-fluoride) is much used because it has a strong pyroelectric effect and can be moulded to any desired form. The journal *Ferroelectrics* had a special issue on PVF and associated polymers in 1981, volume 32, numbers 1, 2, 3, 4. This included 26 papers and a bibliography covering 1961 to 1980. (The bibliography with both subject and author classification contained 654 entries.)

REFERENCES

Barkhausen, H. (1919). 'Zwei mit Hilfe der neuen Verstärker entdeckte Erscheinungen' (two phenomena revealed with the help of the new amplifiers), *Phys. Zeits.*, **20**, 401–403

Bittel, H. (1969). 'Noise of ferromagnetic materials', *IEEE Trans. Mag.*, **MAG-5**, 359–365

Bittel, H. and Lüttgemeier, H. (1963). 'Über spontane zeitliche Schwankungen der Magnetisierung eines Ferromagnetikums' (on the spontaneous time fluctuations of the magnetisation of a ferromagnetic body), *Z. angew. Phys.*, **15**, 476–480

Bittel, H. and Westerboer, I. (1959). 'Kopplungen zwischen Barkhausensprünger als Folge magnetischer Nachwirkung' (coupling between Barkhausen pulses as a consequence of magnetic after-effect), *Ann. Phys.*, **4**, 203–215

Blackburn, H., Robinson, D. A. and Tomlins, G. T. (1975). 'A thermal imaging system using a pyroelectric detector array', *Low Light Level Imaging*, IEE Conference Publication No. 124, pp. 201–206

Bozorth, R. M. (1929). 'Barkhausen effect in Fe, Ni and Permalloy', *Phys. Rev.*, **34**, 772–784

Bozorth, R. M. and Dillinger, J. F. (1932). 'Barkhausen effect III. Nature of changes of magnetization in elementary domains', *Phys. Rev.*, **41**, 345–355

Brophy, J. J. (1958). 'Magnetic fluctuations in molybdenum Permalloy', *J. Appl. Phys.*, **29**, 483–484

Brophy, J. J. (1965). 'Fluctuation in magnetic and dielectric solids', in *Fluctuation Phenomena in Solids* (Ed. R. E. Burgess), Academic Press; New York and London, pp. 1–35

Bukharov, M. B. and Kolachevskii, N. N. (1981). 'Investigation of the correlational relationship between sources of magnetic flicker noise', *Bull. Acad. Sci. USSR, Phys. Series*, **45**, 25–29

Bunkin, F. V. (1956). See *Science Abstracts A* (*Physics*), 1957, 1957 Abstracts 2431

Chen, H. S. (1980). 'Glassy metals', *Rep. Prog. Phys.*, **43**, 353–432

Chynoweth, A. G. (1959). 'Effect of space charge fields on polarisation reversal and the generation of Barkhausen pulses in Barium Titanate', *J. Appl. Phys.*, **30**, 180–285

Deimel, P., Röde, B. and von Trentini, G. (1977). 'Korngrössenabhängigkeit des Barkhausen-Effektes von Eisen bei 300 K und 77 K' (grain size dependence of the Barkhausen effect in iron at 300 K and 77 K), *J. Mag. and Mag. Mat.*, **4**, 235–241

Drokin, A. I., Metlyaev, T. N., Ivanova, A. V., Abelyashev, G. N., Scherbakov, V. N., Shemyakov, A. A. and Chervenchuk, L. P. (1981). 'Effect of magnetostructural changes on the Barkhausen effect in ferrospinels containing cobalt', *Bull. Acad. Sci. USSR, Phys. Ser.*, **45**, 42–45

Eguchi, M. (1925). 'On the permanent electret', *Phil. Mag.*, **49**, 178–192

Fedosov, V. N. and Chursina, E. I. (1981). 'Fluctuational phenomena during magnetic switching of inhomogeneous ferromagnetics', *Bull. Acad. Sci. USSR, Phys. Series*, **45**, 67–71

Fiorillo, F., Mazzetti, P., Vinai, F. and Soardo, G. P. (1977). 'Barkhausen noise in amorphous ferromagnetic materials', *Physica B + C*, **83 B + C**, 803–804

Gemant, A. (1933). *Liquid Dielectrics*, Wiley; New York

Gemant, A. (1935). 'Recent investigations on electrets', *Phil. Mag.*, **20**, 929–952 (Nov. 1935 supplement)

Grachev, A. A. (1950). 'On the discrete/continuous spectrum of induction of a ferromagnetic for cyclic magnetisation' (in Russian), *Dokl. Akad. Nauk SSSR*, **71**, 269–271

Heaviside, O. (1892). *Electrical Papers, Volume 1*, Macmillan; London (see Article 30, section 12, pp. 488–493)

Higham, A. D. and Wilkinson, P. B. (1975). 'Thermal imager using ceramic pyroelectric detectors', *Low Light Level Imaging*, IEE Conference Publication No. 124, pp. 207–211

Hubert, A. (1974). 'Theorie der Domänenwände in geordneten Medien' (theory of domain walls in ordered media), *Lecture Notes in Physics, No. 26*, Springer-Verlag; Berlin, Heidelberg, New York

Keve, E. T., Bye, K. L., Whipps, P. M. and Annis, A. D. (1971). 'Structural inhibition of ferroelectric switching in triglycine sulphate—I. Additives', *Ferroelectrics*, **3**, 39–48

Kittel, C. and Galt, J. (1956). 'Ferromagnetic domain theory', *Solid State Phys.*, **3**, 437–564

Lems, W., Rigners, P. J., Bongers, P. F. and Enz, U. (1948). 'Photomagnetic effects in a chalcogenide spinel', *Phys. Rev. Let.*, **21**, 1643–1645

Lines, M. E. and Glass, A. M. (1977). *Principles and Applications of Ferroelectric and Related Materials*, Clarendon Press; Oxford

Mazzetti, P. and Montalenti, G. (1962). 'Sul calcolo del rumore di fondo in circuiti con elementi utilizzanti materiali ferromagnetici' (on the calculation of the background noise in circuits using elements with components using ferromagnetic materials), *L'Energia Elettrica*, **34**, 562–574

Mazzetti, P. and Montalenti, G. (1963). 'Power spectrum of the Barkhausen noise of various materials', *J. Appl. Phys.*, **34**, 3223–3225

Miller, R. C. (1960). 'On the origin of Barkhausen pulses in $BaTiO_3$', *J. Phys. Chem. Solids*, **17**, 93–100

Néel, L. (1946). 'Principles of a new general theory of the coercive force', *Ann. Univ. Grenoble*, **22**, 299–341

Noble, S. W. and Baxendall, P. J. (1952). 'The design of a practical d.c. amplifier based on the second-harmonic type of magnetic modulator', *Proc. Inst. Elect. Engrs*, **99**, Pt. II, 327–348

Primdahl, F. (1979). 'The fluxgate magnetometer', *J. Phys. E.*, **12**, 241–253

Schroder, K. and Saijja, V. (1982). 'The effects of defects and internal stresses on the magnetic structure of steels', *AIP Conf. Proc. No. 84*, pp. 27–31

Tebble, R. S. (1955). 'The Barkhausen effect', *Proc. Phys. Soc. B*, **68**, 1017–1032

Tebble, R. S. and Newhouse, V. L. (1953). 'The Barkhausen effect in single crystals', *Proc. Phys. Soc. B*, **66**, 633–664

Tebble, R. S. Skidmore, I. C. and Comer, D. W. (1950). 'The Barkhausen effect', *Proc. Phys. Soc. A*, **63**, 739–761

Veselago, V. G., Kutznetsov, V. N. and Makhotkin, V. E. (1981). 'Effect of light on magnetic noise in the magnetic semiconductor $CdCr_2Se_4$', *Bull. Acad. Sci. USSR, Phys. Series*, **45**, 56–58

Warren, K. G. (1961). 'Barkhausen noise in transformer cores', *Electron. Tech.*, **38**, 89–94

Wiegman, N. J. (1977). 'Barkhausen effect in thin films: experimental noise spectra', *J. Appl. Phys.*, **12**, 157–161

Williams, J. and Shockley, W. (1949). 'A simple domain structure in an iron crystal showing a direct correlation with the magnetisation', *Phys. Rev.*, **75**, 178–183

Chapter 4

Hot Carriers and Ballistic Transport: Avalanche Devices and the Gunn Effect

4.1 Introduction

The earliest practical use of the noise of hot electrons was the use of gas-discharge tubes as noise sources in waveguides (Knol, 1951). In this case the electrons were accelerated by the applied field (with intermediate elastic scattering) until they acquired sufficient energy to ionise a gas atom, when they lost this energy by ionisation and inelastic scattering. As a result of the scattering the electrons had a randomness in their velocities; and within the glow-discharge region of the tube their velocities were limited by the value required to produce ionisation and the energy of random velocities was approximately one-fifth of this. In view of the random components of velocity arising from the scattering, the mean-square departure from the drift velocity could be described in terms of an effective temperature which in gas discharges was of the order of 10^4 K; and this could be used as a thermal noise source corresponding to the resistance of the discharge column at an elevated temperature.

Sufficiently high fields to produce hot electrons cannot be applied to metals without producing a current density which would result in a high level of power dissipation and probably cause melting of the metal. But in a semiconductor the current and power levels are much less for a given electric field so that hot electrons may be created, and if current is limited to pulses there need be no appreciable temperature rise. For example, Gasquet et al. (1983) used pulses of 500 ns duration with an interval between pulses of 1 to 10 s to measure a device at 300 K. The limit on electron temperature is the energy required to ionise an atom in the solid. Such ionisation in the solid leads to avalanche effects, which will be discussed later; but for the present only hot electrons (holes) having energy insufficient to ionise atoms will be considered.

4.2 Diffusion noise

The behaviour of hot electrons has become important as a result of the development of small field effect transistors (FET), though experimental measurements may be made on corresponding diode structures, i.e. omitting the gate of the FET. With a device length in the micrometre range the internal electric field (though modified by space charge) can average 10 kV/cm since 1 V/μm equals 10^4 V/cm. If the idea of electron temperature is valid, it should be possible to express the noise as Johnson noise, though one would have to allow for a non-linear V/I characteristic of the device. The Nyquist theorem for thermal noise is directly applicable only as long as the carriers are in thermal equilibrium with the crystal lattice (of which the temperature can be measured), a condition which is not true of hot carriers. But the noise is still due to the random movement of carriers which is superimposed on their drift motion and so may properly be called 'diffusion noise', since diffusion is due to random movements which are independent of drift velocity. When the energy acquired by a carrier from the field is appreciable compared with its thermal energy, the distribution in space of the random velocities is no longer isotropic. This effect in ionised gases was studied by Druyvestein, with whose name the effect in general is associated. In solids the Druyvestein distribution (Druyvestein, 1934) can be expected if $(\mu_0 V/L)(1/u) \gg 1$ where μ_0 is the low-field mobility, u is the velocity of sound in the medium (relevant because only scattering by acoustic phonons is considered), V is the applied voltage and L the length of specimen. Since V/L is the average electric field, this condition is that, had the low-fie.d mobility still applied, the drift velocity would have been greater than the speed of sound in the medium. When the system is in equilibrium one can divide the parameters into components parallel and normal to the electric field and apply the Einstein relation separately to each (Jacobini and Reggiani, 1979):

$$D_{\parallel}(E) = (1/e)\mu_{\parallel}(E)(2/3)\langle \varepsilon \rangle \qquad (4.1a)$$

$$D_{\perp}(E) = (1/e)\mu_{\perp}(E)(2/3)\langle \varepsilon \rangle \qquad (4.1b)$$

$\langle \varepsilon \rangle$ is the mean energy corresponding to random velocities and one hopes that this mean energy can be represented by an equipartition type of formula, $\langle \varepsilon \rangle = (3/2)kT_n$ where T_n is the noise temperature. Gisolf and Zijlstra (1974) calculated that for a uniform planar system the noise temperature is given by

$$\frac{T_n}{T} = 0.58 \frac{\mu_0 V}{L} \left(\frac{1}{u} \right) \qquad (4.2)$$

This result was confirmed by Van Vliet et al. (1975), who examined the ohmic, space-charge-limited and trapping regimes; and they stated that the salami method of analysing a non-uniform system would give a correct result only if

$$\int_0^L dx[E_0(L) - 2E_0(x)]\mu k T_e(x)n(x) = 0 \qquad (4.3)$$

Recent work has concentrated on lower than room temperature, e.g. 77 K, where generation-recombination noise is evident as well as diffusion noise and experimental results have been shown in terms of $T_n - T$ rather than T_n/T. Gasquet, Tijani et al. (1983) reported for n-silicon at 295 K a noise temperature which saturated at $T_n - T \simeq$ 180 K with a field of 10^4 V/cm, while Gasquet, Fadel and Nougier (1983) found $T_n - T$ nearer 500 K in n-InP at 300 K and 10^4 V/cm. (The latter was a high-frequency asymptote since g-r noise was dominant below 10^9 Hz.)

The description of the energy of the electrons (holes) by a temperature is strictly valid only if their random velocities (1) are isotropic and (2) have a Maxwell–Boltzmann distribution. In the solid state the degree of isotropy may be affected both by differential scattering in the electric field (Druyvestein) and by a phenomenon known as 'inter-valley scattering'. The 'valleys' refer to the shape of energy curves plotted in k space, but for the present purpose it suffices to regard the different valleys as different modes of propagation, often having different effective mass of carriers and different mobility. This extent of anisotropy is allowed for in Equations (4.1a) and (4.1b). Condition (2) was tested in a Monte Carlo simulation by Hill et al. (1979) and they found that the distribution of velocities parallel to the electric field was sufficiently close to Maxwell–Boltzmann to justify the concept of 'longitudinal temperature'.

Zijlstra (1978) derived the following general formula for the thermal and $1/f$ noise in a hot-electron device, of length L, sufficient for the concept of diffusion to be applicable:

$$S_{\Delta v}(f) = \frac{4\varepsilon q A}{I_2^0} \int_0^{E_L} \frac{D(E)(E_L - E)^2 \, dE}{(1 - qAp_aE\mu(E)/I_0)^3}$$

$$+ \frac{\varepsilon q A}{f I_0^2} \int_0^{E_L} \frac{\alpha(E)(E_L - E)^2 \, dE}{(1 - qAp_aE\mu(E)/I_0)^3} \qquad (4.4)$$

where p_a is density of ionised acceptors, I_0 is steady current, E_L is electric field at the collecting terminal, A is area of cross-section, ε is the permittivity of the material and $-q$ is the electron charge. This requires that one first determine the functions relating D, μ and α to E, and then perform the integration over the range of E which exists in the device. Bosman et al. (1981) observed the noise from a $p^+\pi p^+$

planar silicon device, with boron doping, of length 40 μm. From values of $1/f$ noise at 78 K they concluded that in relation to $1/f$ noise the carriers became 'hot' (in the sense that α began to decrease rapidly from its low-field value) when the velocity of light holes was equal to the velocity of sound in the material. From this they concluded that $1/f$ noise is associated with scattering by acoustic phonons. (The mobility was only slightly reduced at this field strength.) They also found that g-r noise was significant at frequencies below 10^6 Hz and identified contributions from six different levels of impurity traps.

The simplistic idea that an electron will have an unhindered trajectory (ballistic transport) if the device length is less than the mean free path is inadequate because most electrons will have a flight length less than the *mean* free path (Cook and Frey, 1981) so that there is no drastic change in characteristic as the device length is reduced through that of the mean free path. Scattering may be regarded as a frictional force which retards the motion of electrons under the applied field; reduction of scattering reduces this force but does not define a minimum length over which scattering can be ignored. 'Ballistic transport' is therefore not to be expected to show dramatic effects at a particular length of FET.

4.3 Monte Carlo simulation

Most theoretical investigations of very short devices use Monte Carlo methods which avoid the idea of a mean free path common to all electrons. In a Monte Carlo simulation on a digital computer one traces the history of one or more particles by using random number generators to select for each, at a given moment in time, the value of free time before the next collision, the initial velocity and the speed and direction after collision. The model is advanced by small steps in time, each new step starting with the positions and velocities established in the previous step. One can either take a single electron and average over many transits through the simulated device or take the ensemble average over a number of electrons travelling simultaneously; but in either case one must at each step use the Poisson equation to take account of the effect of space charge on the local electric field. (For details see Zimmerman and Constant, 1980.) From either method one can deduce the average $I - V$ characteristic for comparison with that calculated by methods depending on the bulk characteristics of the device. From the multi-electron method one can also calculate a noise temperature from the mean-square fluctuation of velocity. This simple model assumes that there is no correlation between individual electrons. Such correlation may exist over times in the picosecond region, so correction for correlation effects is necessary at the highest frequency (Moglestue, 1983). Since the noise (temperature) varies along the length of the conductor the noise at the terminals should be calculated by the field-impedance method, which is more rigorous than the salami method.

The difference between noise temperature T_n and ambient temperature T_0 which may be found in practice is illustrated in Fig. 4.1, (a) for n-type silicon and (b) for n-type InP, both at 360 K. The component of noise which is independent of field is attributed to generation-recombination noise which is significant only below 500 MHz. It is possible to deduce the variation of diffusion coefficient with field strength from the characteristic of noise temperature versus field strength at a frequency of 10.5 GHz, where g-r noise is negligible, and Fig. 4.2 shows that this derivation of diffusion coefficient is in good agreement with that found by a Monte Carlo simulation of the electron transport carried out by Hill et al., 1979. These results refer to indium phosphide at 300 K. Similar results for silicon have been reported by Gasquet, Tijani et al., but mostly at the lower temperature of 77 K.

A simpler method of Monte Carlo simulation for FETs has been proposed by Cappy (Cappy, 1981; Carnez et al., 1981), using carrier transport equations of momentum and energy relaxation obtained by integration of the Boltzmann equation; and this has been applied to a dual-gate MOSFET having sub-micrometre lengths of gate (Allamando et al., 1983). The theoretical prediction for a particular device is a linear relation between noise figure in decibels and frequency in GHz, ranging from 2 dB at 8 GHz approximately to 4.5 dB at 18 GHz; and measured values of noise are in good agreement with this.

An unusual application of hot electrons is in the cryogenic InSb bolometer which was first proposed by Rollin (1961) for the detection of millimetre and sub-millimetre radiation. The principle is that the conduction electrons are heated (above lattice temperature) by incident radiation and this produces a reduction in mobility which can be observed as an increase of resistance. The mobility varies as $E^{-1/2}(m^*)^{-3/2}$ and InSb is an especially suitable material because the effective mass m^* of an electron in its conduction band is only 0.015 times the mass of a free electron. The various sources of noise associated with this device will be detailed in Chapter 7.

4.4 Submicron devices

If the length of the device can ultimately be reduced to something comparable with the mean free path in the material, an electron will pass straight through, without its energy being 'thermalised' by

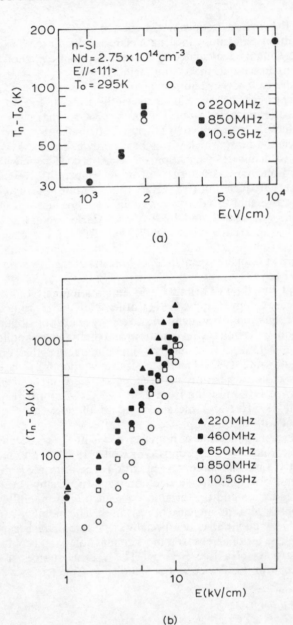

Fig. 4.1 Excess noise temperature vs. electric field at various frequencies (a) in n-type silicon at 300 K (after Gasquet, Tijani et al., 1983) and (b) in n-type InP at 300 K (from Gasquet, Fadel and Nougier, 1983)

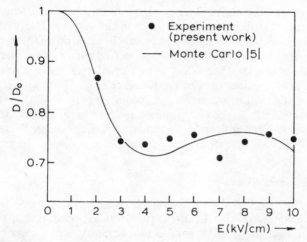

Fig. 4.2 Diffusion coefficient vs. electric field in n-type InP at 300 K: comparison of experiment (●), with Monte Carlo simulation (——). The low-field diffusion coefficient is $D_0 = 117 \; cm^2/s$ (Gasquet, Fadel and Nougier, 1983)

scattering. Such a mode of transport is known as 'ballistic' and should produce shot noise with the appropriate correction to the basic shot-noise formula if the transit time is comparable with the period of the working frequency. (See Appendix I for shot noise.) It is not yet possible to manufacture such small devices, though some silicon devices have an active length (e.g. length of gate in a FET) of a quarter of a micrometre. A nearer approximation to ballistic transport can be expected in gallium arsenide, because the mean free path in it is greater than in silicon. Because the crystal of GaAs is polar, whereas a silicon crystal is covalent, it does not have the same feature of a limiting drift velocity; and the field strength which can be applied to GaAs is limited by the onset of negative differential mobility (see below in relation to the Gunn effect), whereas the limit in Si is the onset of avalanching. Schmidt et al. (1983) quote the mean free path in GaAs at room temperature as effectively 0.1 μm for electrons of 0.05 eV energy and 0.2 μm for electrons of 0.5 eV. An electron passing through a device of active length 0.4 μm should then on average be scattered two to four times; and this can be called 'near ballistic' transit—there should certainly be no question of the electron being 'thermalised' during such a transit. Schmidt et al. experimented on five-fold stacks, either alternately p and n^+ layers with a thickness of 0.47 μm per layer or n and n^+ with a thickness of 0.4 μm per layer, and calculated the $I - V$ characteristic from a generalised Fry model (Fry, 1921), which accounts for space charge for permittivity and for injection into the active layer of electrons which have a Maxwell–Boltzmann distribution of velocities in the n^+ regions. The predicted $V-I$ characteristic was confirmed exactly by experiment for the n^+nn system, but for the n^+pn^+ system the observed current was more than a decade below the theoretical value for voltages below about one volt, both at a temperature of 300 K and at 77 K. The observed noise was more difficult to reconcile with theory, since it consisted of a low level of noise of $1/f$ type plus white noise which the authors described as 'thermal (-like)' and which was equal to Nyquist noise. There is as yet no clear evidence on ballistic transport as a distinct mode.

4.5 The Gunn effect

The Gunn effect is a hot-electron phenomenon associated with a particular material characteristic. Its practical importance is that it has been used to produce solid state oscillators with power in the range 0.2 to 100 W in the frequency range 5 to 50 GHz. In crystals with polar valence bonds, e.g. in GaAs, the electron velocity decreases sharply with increasing field beyond a threshold value (between 3 and 5 kV/cm in GaAs) as a result of scattering of electrons out of the valley of lowest energy into satellite valleys of higher energy in which the effective mass is greater and therefore mobility less. (The term 'valley' comes from the shape of the wave number versus energy graphs.) This effect is not found in silicon because the valence bonds are covalent and increasing field only results in a gradual decrease in mobility as the electron drift velocity approaches a terminal value. Any conductor which exhibits a negative differential mobility (velocity decreasing as field increases over a certain range) will from a circuit point of view show a negative differential resistance and therefore can in principle be used either as an amplifier or as an oscillator. If the decrease in conductivity, corresponding to the decrease in velocity, occurs in part only of the device, with the current through the whole remaining constant, there must be an enhancement of electric field in this region which may, in turn, enhance the effect. In the Gunn oscillator this region is known as a domain and travels through the device with a velocity equal to the drift velocity of electrons outside the domain. It is possible for several domains to exist simultaneously within the length of the device; but the preferred mode is to have only one domain at a time, which is initiated at or near the cathode and travels to the anode where it is discharged, whereupon a new domain is initiated to follow the same cycle. The oscillatory behaviour depends on the extent to which the oscillator voltage, added to the bias voltage, can govern the formation of domains; and a major factor in this is the quantity Z_0/R_0, where R_0 is the magnitude of negative resistance presented by the device to a circuit and Z_0^2 is the L/C ratio of the circuit, including the internal capacitance of the device. For $Z_0/R_0 \leqslant 2$ the only possible mode of oscillation is that governed by the transit time of a domain from cathode to anode under the bias field. (This was the original mode of working (Gunn, 1963).) For higher ratios of Z_0/R_0 the oscillatory voltage may be sufficient to control the initiation and quenching of domains; and provided the drift time $T_D \gtrsim \pi\sqrt{LC}$, so that the domain may be quenched by the decreasing overall field before it has reached the anode, the oscillatory frequency is controlled by the circuit. This is the preferable mode of operation.

There are three sources of noise in a Gunn effect device:

(1) The electrons in the whole of the conduction process are 'hot', having an effective temperature of about 700 K.

(2) The exchange within the domain of electrons transferring between the main valley and the satellite valleys (intervalley scattering) is a source of noise. This intervalley scattering within the domain is equivalent to a current which circulates within the domain, not round the whole circuit; and its effect on the external circuit therefore needs to be calculated by the impedance field method to take account of a noise current which extends over only a part of the conduction path between the terminals of the device. The intervalley scattering noise is greatest when the electrons are about equally divided between the lowest valley and the others, because if the electrons are mostly in one state or the other there is little intervalley scattering. This is the condition of greatest negative differential mobility, as illustrated in Fig. 4.3 (Chatterjee and Das, 1983), and hence the greatest negative resistance, i.e. the condition which one is likely to want to use. If the working frequency is not much over 10 GHz the effective total noise will be that of the low-frequency plateau in Fig. 4.4, i.e. more than 10 dB above thermal equilibrium noise. The intervalley scattering is a slower process than the thermal scattering within a valley, so the two processes are uncorrelated and their contributions to the total noise are additive.

(3) The time taken to initiate a domain is subject to statistical fluctuation and the consequent jitter in starting time appears as frequency-modulation noise (Meade, 1972). One method of stabilising the creation of domains is to introduce a local enhancement of the electric field through a singularity or 'notch' in the doping concentration near the cathode, e.g. by ion bombardment. In one case (Howes et al., 1979) the noise at frequencies less than 1 kHz from the carrier was reduced, but at the cost of increasing the

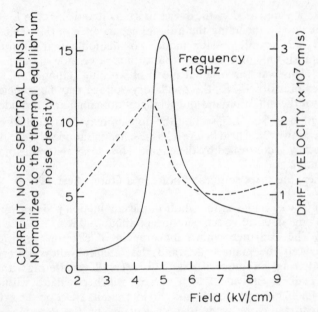

Fig. 4.3 Calculated drift velocity (– – –) and low-frequency noise (——) vs. electric field in GaAs (from Chatterjee and Das, 1983)

noise further out. (See Chapter 6 on Oscillator Noise for the effect of circuit Q on phase modulation and on amplitude modulation; and for other factors in the noise performance of Gunn oscillators.)

4.6 Avalanche multiplication

One ought first to ask why an avalanche in a semiconductor can produce stable amplification, while an arc between metal electrodes is a form of avalanche which tends to grow without limit and is therefore destructive. There are, in fact, three reasons:

(1) The impact of positive gas ions on the cathode plays an important part in the metallic arc, but there is no corresponding electrode effect in the semiconductor avalanche.
(2) The ionisation coefficients of electrons and of holes usually differ—if they were equal and the device were symmetrical one could have a catastrophic growth of the avalanche.
(3) The probability that an impact will cause ionisation is a function of the local electric field, as well as of the energy of the impacting electron or hole.

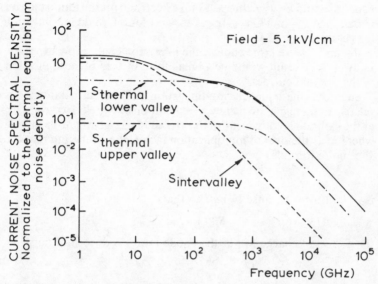

Fig. 4.4 Calculated components of noise (spectral intensity of noise current) in GaAs biased to the region of negative differential mobility, and their variation with frequency (from Chatterjee and Das, 1983)

Apart from the inherent difference in ionisation coefficients, the relative effectiveness of holes and electrons can be varied by making the electric field non-uniform, e.g. by non-uniform doping.

If the electron energy is sufficiently increased, the impact of an electron on an atom may cause ionisation, i.e. cause the release of an additional electron. In semiconductor terminology this is described as the creation of a hole-electron pair, since the vacancy left by the removal of an electron from an atom is regarded as a hole which can move in the opposite direction under the applied field. At this point it should perhaps be remembered that a 'hole' is a useful concept but not a physical particle. When a hole is created by the removal of a previously bound electron and the hole is said to move down the electric field, what happens is that a whole series of other bound electrons are able to move in turn up the field with little net expenditure of energy, by a kind of 'musical chairs' process, so that the vacancy moves down the field. It is therefore not surprising that the ionisation coefficient for holes is not the same as that for electrons. The coefficients give the probability per unit length of path traversed that ionisation will be produced; and some values of the coefficients in different materials have been collected by Sze (1981) from various sources. If α is the coefficient of ionisation per unit length for electrons,

β for holes and $k = \beta/\alpha$, then $k < 1$ for silicon and for the ternary alloys $GaAs_{0.88}Sb_{0.12}$ and $Ga_{0.45}In_{0.55}As$, $k = 1$ for SiC and GaP but $k > 1$ for germanium and GaAs.

If α and β were large enough the ionisation could grow exponentially and the result would be a breakdown instead of multiplication of the original current.*

So long as the avalanche multiplication is effectively instantaneous relative to the signal frequency, the liberation of one electron charge q at the cathode is equivalent to the arrival of charge Mq at the anode where M is the overall multiplication factor. Remembering that the shot noise formula

$$S_i(f) = 2qi$$

is of that form because $i = nq/t$ so that

$$S_i(f) = 2nq^2/t$$

the arrival of charge Mq at the anode must produce shot noise

$$S_i(f) = 2M^2q^2n/t = 2M^2qi$$

But M is the *average* multiplication and there is also statistical fluctuation at each of the stages of ionisation which together account for M. This additional fluctuation may be represented by a factor F, or in an avalanche photodiode by F_s for the signal current and F_d for dark current which may not be the same. If the ionisation process is regarded as continuous, the following formula (which is usually attributed to McIntyre) gives the signal/noise ratio at the output of an avalanche photodiode (Webb *et al.*, 1974):

$$\frac{S}{N} = \frac{(i_s M)^2}{2qi_{ds} + (i_s F_s + i_{db} F_d)M^2 B + i_{na}^2} \tag{4.5}$$

(i_{na}^2 represents the noise from the bias resistor and the input noise of the following amplifier and i_s is signal input current†). The assumption that the ionisation process can be represented by a continuous function causes the noise to be over-estimated to an extent which never exceeds 5% and is negligible if the number of ionisations per initial electron exceeds 8. If one treats each ionisation as a separate event it has been shown theoretically (Van Vliet *et al.*, 1979) and confirmed experimentally (Gong and Van Vliet, 1981) that the mean multiplication factor is

---

* For a quantitative treatment of the condition for breakdown see Shy Wang, 1966. When electrons and holes in a plasma have equal ionising capabilities the resulting instability is the mechanism of the IMPATT oscillator.

† See Chapter 7, Radiation Detectors, for further discussion of this formula.

$$M = \frac{(1+\lambda)^N(1-k)}{(1+k\lambda)^{N+1} - k(1+\lambda)^{N+1}} \qquad (4.6)$$

where λ is the *a priori* probability that an electron which has travelled in the electric field a distance sufficient to gain the energy necessary for it to cause ionisation, μ is the corresponding probability for a hole and $k = \mu/\lambda$. Note that these quantities are not the same as the α and β used earlier which were continuous probabilities per unit distance that ionisation would occur and were functions of electron energy and of field. The variance of the instantaneous value of multiplication after N ionisations is

$$\text{var } X_N = \frac{M(M-1)(1-k)}{2+\lambda+k\lambda} \left\{ -\lambda + 2\frac{1-k\lambda^2}{1+k\lambda} \left[Mk\frac{1+\lambda}{1-k} + \frac{1}{1+\lambda} \right] \right\}$$

$$(4.7)$$

The output noise can be represented by a noise-diode (shot noise) equivalent current which is related to the primary current by

$$I_{eq}/I_{pr} = M^2 + \text{var } X = FM^2 \qquad (4.8)$$

where F is an excess noise factor, $F = 1 + (\text{var } X)/M^2$. If M tends to infinity, and therefore is much larger than unity and than the parameters k and λ, var X_N in (4.7) tends to M^3; and this, in turn, will predominate over M^2 in (4.8). Thus M^3 is an upper bound for I_{eq}/I_{pr} when M is very large.

REFERENCES

Allamando, E., Salmer, G., Constant, E., Radhy, N. E., Cappy, A. and Carnez, B. (1983). 'A new noise model of submicrometer dual gate MESFET', *Noise in Physical Systems and 1/f Noise* (Ed. M. Savelli, G. Lecoy and J-P. Nougier), North-Holland; Amsterdam, pp. 177–179

Bosman, G., Zijlstra, R. J. J. and van Rheenen, A. (1981). 'Hot carrier transport noise in p-type silicon', *Sixth International Conference on Noise in Physical Systems*, pp. 409–413

Cappy, A. (1981). 'Sur un nouveau modèle de transistor à effet de champ à grille submicronique', *Thesis, Lille University*

Carnez, B., Cappy, A., Fauquembergue, R., Constant, E. and Salmer, G. (1981). 'Noise modeling in submicrometer-gate FET's', *IEEE Trans Electron. Dev.*, **ED-28**, 784–789

Chatterjee, A. and Das, P. (1983). 'Current noise in multivalley semiconductors', *Noise in Physical Systems and 1/f Noise* (Ed. M. Savelli, G. Lecoy and J-P. Nougier), North-Holland; Amsterdam, pp. 161–164

Cook, R. and Frey, J. (1981). 'Diffusion effects and "ballistic transport"', *IEEE Trans Electron. Dev.*, **ED-28**, 167–169

Druyvestein, M. J. (1934). 'Bemerkungen zu zwei früheren Arbeiten über die Elektronendiffusion' (observations on two earlier papers on electron diffusion), *Physica*, **1**, 1003–1006

Fry, T. C. (1921). 'The thermionic current between parallel plane electrodes; velocities of emission distributed according to Maxwell's law', *Phys. Rev.*, **17**, 441–452

Gasquet, D., Fadel, M. and Nougier, J-P. (1983). 'Noise of hot electrons in indium phosphide', *Noise in Physical Systems and 1/f Noise* (Ed. M. Savelli, G. Lecoy and J-P. Nougier), North-Holland; Amsterdam, pp. 169–171

Gasquet, D., Tijani, H., Nougier, J-P. and van der Ziel, A. (1983). 'Diffusion and generation recombination of hot electrons in silicon at 77 K', *Noise in Physical Systems and 1/f Noise* (Ed. M. Savelli, G. Lecoy and J-P. Nougier), North-Holland; Amsterdam, pp. 165–167

Gisolf, A. and Zijlstra, R. J. J. (1974). 'Noise and hot carrier effects in a single injection solid state diode', *Solid State Electron.*, **17**, 839–841

Gong, J. and Van Vliet, K. M. (1981). 'Noise measurements on photo avalanche diodes', *Sixth International Conference on Noise in Physical Systems*, pp. 113–117

Gunn, J. B. (1963). 'Microwave oscillations of current in III–V semiconductors', *Solid State Commun.*, **1**, 88–91

Hill, G., Robson, P. N. and Fawcett, W. (1979). 'Diffusion and the power spectral density of velocity fluctuations for electrons in InP by Monte Carlo methods', *J. Appl. Phys.*, **50**, 356–360

Howes, M. J., Morgan, D. V., Blundell, R. and Greenhalgh, S. (1979). 'Noise reduction in transferred electron modules', *IEEE Trans. Electron. Dev.*, **ED-26**, 237–238

Jacobini, C. and Reggiani, L. (1979). 'Bulk hot-electron properties of cubic semiconductors', *Adv. Phys.*, **28**, 493–553

Knol, K. S. (1951). 'Determination of the electron temperature in gas discharges by noise measurements', *Philips Res. Rep.*, **6**, 123–126

Meade, M. L. (1972). 'Investigation of noise in Gunn effect oscillators', *Thesis, University of Reading*

Moglestue, C. (1983). 'Monte Carlo particle modelling of noise in semiconductors', *Noise in Physical Systems and 1/f Noise* (Ed. M. Savelli, G. Lecoy and J-P. Nougier), North-Holland; Amsterdam, pp. 23–25

Rollin, B. V. (1961). 'Detection of millimetre and sub-millimetre wave radiation by free carrier absorption in a semiconductor', *Proc. Phys. Soc.*, **77**, 1102–1103

Schmidt, R. R., Bosman, G., Van Vliet, C. M. and van der Ziel, A. (1983). 'Noise in near-ballistic n^+nn^+ and n^+pn^+ gallium arsenide submicron diodes', *Noise in Physical Systems and 1/f Noise* (Ed. M. Savelli, G. Lecoy and J-P. Nougier), North-Holland; Amsterdam, pp. 173–176

Shy Wang (1966). *Solid State Electronics*, McGraw-Hill; New York

Sze, S. M. (1981). *Physics of Semiconductor Devices* (2nd edn), Wiley; New York

Van Vliet, K. M., Rucker, L. M. and Friedman, A. (1979). 'Theory of carrier multiplication and noise in avalanche devices—Part II: Two-carrier', *IEEE Trans. Electron. Dev.*, **ED-26**, 752–764

Van Vliet, K. M., Friedman, A., Zijlstra, R. J. J., Gisolf, A. and van der Ziel, A. (1975). 'Noise in single injection diodes: II, Applications', *J. Appl. Phys.*, **46**, 1814–1823

Webb, P. P., McIntyre, R. J. and Conradi, J. (1974). 'Properties of avalanche photodiodes', *RCA Rev.*, **35**, 234–278

Zijlstra, R. J. J. (1978). 'Nonohmic transport fluctuations in semiconductors', in *Noise in Physical Systems* (Ed. D. Wolf), Springer-Verlag; Berlin, pp. 90–95

Zimmerman, J. and Constant, E. (1980). 'Application of Monte Carlo techniques to hot carrier diffusion noise calculation in unipolar semiconducting components', *Solid State Electron.*, **23**, 915–925

Chapter 5

Cryogenic Devices

5.1 Introduction

Since the most ubiquitous and apparently inevitable form of noise is thermal or Johnson noise, a natural step is to lower the temperature. This is fully effective only if the source of signal is also at a low temperature and this applies to the background noise from an antenna used in radio astronomy or in the earth station of a satellite communication system. (See Section 1.9, 'The temperature of radiation resistance'.) Refrigeration would clearly be impossible with thermionic devices (in which noise is related to cathode temperature), and any solid state device depending on thermal excitation would have its characteristics severely modified by drastic cooling. The seven types of device which will be examined in this chapter are the maser, the parametric amplifier, the tunnel diode, the SIS mixer, the Super-Schottky diode and cooled GaAs FETs for microwave and millimetre-wave use; and the Josephson junction incorporated in the SQUID for measurement of very small quantities at low frequencies.

Several of the cryogenic devices (cavity masers, cavity parametric amplifiers, tunnel diodes) are negative-resistance amplifiers and in the simplest case they are merely shunted across the signal source so as to cancel most of the resistance of the source and of the load to which it is connected. But in this arrangement the output of the amplifier is connected to its input, which has two disadvantages:

(1) It is potentially unstable and high gain requires working near the threshold of instability.
(2) Any noise (e.g. Johnson noise) arising in the output circuit is applied to the input and is amplified.

It is therefore usually considered essential to use a circulator. This has a minimum of three ports but a four-port circulator may be preferred, as shown in Fig. 5.1, so that noise originating in the output circuit is not fed back to the signal source. The difficulty immediately arises that the ferrite commonly used in microwave circulators cannot be cooled to a low temperature; and the inevitable losses at room

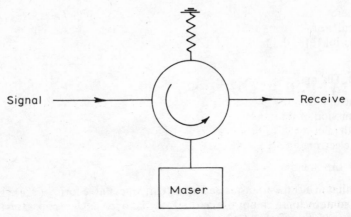

Fig. 5.1 Arrangement of a 4-port circulator

temperature make a relatively large contribution to the total noise associated with a cooled system. Ditchfield (1968) gave the following example of the components of the effective noise temperature of a cavity maser amplifier with a circulator:

$$T_{\text{eff}} = \underset{\substack{\text{input} \\ \text{circuit}}}{27.8} + \underset{\substack{\text{maser} \\ \text{material}}}{3.8} + \underset{\substack{\text{cavity} \\ \text{walls}}}{0.6} = 37.7\,\text{K}^{\dagger}$$

The input circuit, including a circulator, has here increased the total noise by a factor of more than eight times. Later, it was found possible to cool a suitable ferrite, either nickel-zinc or lithium-zinc, to 17 K where the insertion loss of a circulator using it was 0.7 dB at 46 GHz (Edrich *et al.*, 1973). Alternatively, a semiconductor such as InSb with a very long electron free path at low temperature becomes anisotropic in a magnetic field (owing to Hall effect and magnetoresistance) and can be used to make a low-temperature circulator without restriction on the lowest temperature attainable.

The performance of a room-temperature amplifier is usually defined by the noise figure F which is the ratio of output (noise/signal) to input (noise/signal), i.e. the deterioration of signal/noise resulting from passing the signal through the device in question. It may be expressed primarily as a numerical ratio, though this is often converted to decibels. But the overall noise performance of a system including a refrigerated amplifier is better described by a noise temperature T_n (or effective temperature T_{eff}) and the relationship between the two is

$$T_n = (F-1)T_a \tag{5.1}$$

† This is a weighted combination of components.

where T_a is the ambient temperature relevant to the device. Thus an ideal noiseless amplifier* would have $F = 1$ and $T_n = 0$; and it is possible for an amplifier with $F < 2$ ($F < 3$ dB) to have $T_n < T_a$.

5.2 The maser

The operation of the maser (Microwave Amplification by Stimulated Emission of Radiation) is the same as that of the now-familiar laser, with microwave radiation substituted for light radiation. The number of electrons in an energy state hv would normally be

$$n = n_0 \exp(-hv/kT)$$

so that in a three-level device $n_3 < n_2 < n_1$ when $hv_3 > hv_2 > hv_1$; but if by some means it can be brought about that n_3 is greater than n_2 or n_1 the population is said to be inverted. The spontaneous or thermal transfers between states must be weighted so as to preserve the normal distribution with fewer electrons in the higher level. If p_{ij} is the probability of transfer from state i to state j, this requires

$$\frac{p_{ij}}{p_{ji}} = \exp(-\Delta E_{ij}/kT)$$

But if radiation at frequency corresponding to the difference

$$\Delta E_{ij} = hv_j - hv_i$$

is incident on the system it will stimulate transitions between the two states *equally in either direction*. In the normal condition, with more electrons in the lower level, the incident radiation will be mainly absorbed in raising electrons to the higher level; but with an inverted population most of the transitions will be *downward*, with emission of radiation, producing a negative-resistance effect. Random noise corresponding to spontaneous transitions between the two states is equivalent to thermal noise in the effective negative resistance at the negative temperature which would produce the inverted populations (Pound, 1957). This temperature is approximately

$$T_m = T_0 \Delta n_0 / \Delta_n$$

where T_0 is the ambient temperature, Δn_0, Δn are the population differences between the two levels with and without inversion, Δn_0 being negative, so that T_m is negative. The combination of negative temperature with negative resistance produces a positive value for thermal noise, according to the Nyquist formula, and this represents the noise arising from spontaneous transitions. The practical require-

---

* But see Section 1.7.

ment is to find a system in which energy can be continuously fed in at one frequency in order to maintain the inverted population while energy is taken out, through the negative resistance effect, at a different frequency. A typical system is the three-level maser in which energy is fed in at the 'pump' frequency corresponding to excitation from level 1 to level 3, the spontaneous transition to intermediate level 2 ensures an excess population at that level, i.e. an inversion between levels 1 and 2, and the signal frequency corresponds to the difference in energy between levels 1 and 2. Details of the dynamics of a three-level system have been given by Schultz-Dubois *et al.* (1959). In such a case the pump frequency must be higher than the signal frequency, corresponding to the energy of the transition $1 \rightarrow 3$ being greater than that of $1 \rightarrow 2$. As an extreme case, the pump frequency may be optical in a maser used for microwave amplification.

The simple maser outlined above presents two difficulties. The first is that the signal frequency must match a frequency which is characteristic of the material used to construct the maser. This can be partially overcome if the frequency can be varied somewhat by applying a magnetic field. With 'synthetic ruby' (0.035% of Cr^{+++} in Al_2O_3) fields up to 40,000 A/m have been used. Another material, which has the additional feature of a high dielectric constant which may be useful in travelling wave masers, is rutile doped with ferric iron, Fe^{+++} in TiO_2. The other difficulty is that the maser as a negative resistance is a one-port device with which input and output are connected to the same pair of terminals. This is potentially unstable, the amount of amplification depending on the exactness of match between the negative conductance of the maser and the positive conductance of the external circuit; and it also means that any noise generated in the output circuit will be amplified by the maser, as well as noise associated with the signal. This difficulty can be overcome by inserting a circulator between the maser and the rest of the system so as to separate the output from the input; but the circulator must involve some additional loss and resistive component which will deteriorate the signal/noise ratio.

This second difficulty is eliminated in the travelling wave maser. This uses a slow-wave structure which can support both signal and pump frequencies and which incorporates maser material in order to provide gain. 'Tuning' is again accomplished by varying a magnetic field. As the natural bandwidth of a maser is that of a spectral line which is small—say, 30 MHz in 6 GHz upwards, or 0.5% or less—the bandwidth of a travelling wave maser is sometimes extended by changing the magnetic field over parts of the length. There may be a single abrupt change, in which case it is called a stepped field, and the effect is the same as stagger tuning in a chain of independent resonant circuits. Ferrite or equivalent material may be incorporated in the travelling wave structure, to ensure strictly unilateral action. The maser is usually cooled by liquid helium to 4 K or lower and the noise temperature of a complete maser amplifier, including the contribution from losses in the input and output circuits, may be 10 to 15 K.

Carter *et al.* (1969) quote a noise temperature of 8 ± 1 K for a maser at Goonhilly earth station for satellite communication, while de Grasse and Scovill (1960) measured the noise temperature at 5.8 GHz of a preamplifier incorporating a ruby travelling-wave maser cooled to 1.6 K and found a value of 10.7 ± 2.28 K.

Ditchfield (1968) pointed out that theoretically it would be possible at 4 GHz and 2 K to detect a signal consisting of about 10 photons in a time equal to the reciprocal of bandwidth. When $hf > kT$ it is logical to think of photon detection rather than energy detection, but noise temperatures are still used as a comparative measure.

5.3 Parametric amplifiers

The form of parametric amplifier which is simplest to appreciate consists of a resonant LC circuit in which the capacitance can be varied at twice the resonant frequency. If the capacitance is reduced every time the charge on it is a maximum, work will be done by the mechanism causing the capacity change and the oscillatory energy in the resonant circuit will be amplified. This arrangement suffers from the defect that the phasing of the capacitance change, relative to the oscillatory voltage, is critical. But amplification can still be obtained if the 'pump' frequency is not equal to the frequency of the signal; and the theoretical foundation of the general parametric amplifier is found in the Manley and Rowe relations, which were originally produced in relation to modulators and demodulators (Manley and Rowe, 1956).

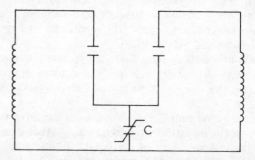

Fig. 5.2 Schematic circuit of variable-capacitance parametric amplifier. C is a non-linear capacitance which is varied at 'pump' frequency

If two frequencies are applied simultaneously to a non-linear reactance (e.g. two sinusoidal voltages applied to a non-linear capacity) having an arbitrary form of non-linearity, then, in general, all sums and differences of multiples of the input frequencies will be produced. The reactance cannot dissipate power and the two key equations are

$$\sum_{n=0}^{\infty} \sum_{m=-\infty}^{\infty} \frac{m W_{mn}}{m f_1 + n f_0} = 0 \tag{5.2}$$

$$\sum_{m=-\infty}^{\infty} \sum_{n=0}^{\infty} \frac{n W_{mn}}{m f_1 + n f_0} = 0 \tag{5.3}$$

where f_1 and f_0 are the two input frequencies and W_{mn} is the power at a frequency equal to m times f_1 plus n times f_0. Some powers must be positive and some negative if the summations are to vanish; and if only three frequencies are involved, namely 'pump' which largely controls the non-linearity, signal and 'idler' which is the difference between the other two, the power in to the non-linear reactor is positive at the pump frequency and negative at the other two. Figure 5.1 illustrates the scheme for a non-linear capacitance (usually a varactor diode) coupling two tank circuits resonant at the signal frequency f_1 and the idler frequency f_2 respectively. The negative power at f_1 and f_2 means that both these circuits will in effect have a negative conductance superimposed on the positive conductances which represent their various losses; and this will amplify the signal initially present in the first circuit and increase the output of the difference frequency, proportional to signal, in the idler circuit. Note that the Manley and Rowe Equations (5.2), (5.3), show that the power at each frequency depends on the relative frequencies and on the form of the non-linearity only in so far as that determines which frequencies are present with appreciable power and so contribute to the summations. But if one wants to work with voltages or admittances one must use the characteristics of the signal and idler circuits as was done by Heffner and Wade (1958).

If a varactor diode is used the first requirement is to ensure that the diode is always sufficiently reverse biased to prevent the flow of current, since any current through the diode would generate shot noise.

If the parametric amplifier is used as a negative conductance applied across the signal circuit at frequency f_1, then denoting by G_1, G_s and G the conductances of the input circuit, signal source and magnitude of negative conductance induced by the parametric operation, the noise figure is

$$F = 1 + \frac{G_1}{G_s} + \frac{G}{G_s}\frac{f_1}{f_2} + \varphi_1(C) + \varphi_2(C) \qquad (5.4)$$

where $\varphi_1(C)$ and $\varphi_2(C)$ represent random fluctuations arising within the non-linear capacitance and are detailed by Heffner and Wade. The series resistance of a varactor diode can be incorporated in the loss resistances of the tuned circuits; and other sources of fluctuations within the non-linear capacitance should be negligible in the varactor diode, though the corresponding effects may be appreciable in an electron-beam parametric amplifier. Thus thermal noise in the loss components of the tuned circuits is the dominant factor in a solid state parametric amplifier, and can be reduced by cooling. Heffner and Wade omit the noise generated in the load conductance, saying in their Appendix II 'The noise emanating from the load conductance is not held against the amplifier'. But, as already mentioned in connection with the maser, one of the disadvantages of negative-resistance amplifiers is that any noise emanating from the load is amplified with the signal, which it would not be in any form of unilateral amplifier. A circulator is necessary to avoid this effect and a component must be added to represent circulator loss. The gain at resonance in this negative-resistance mode is

$$\text{Power gain} = \frac{4 G_s G_L}{(G_{T1} - G)^2} \qquad (5.5)$$

where G_s, G_L, G_{T1} and G are the conductance of the signal source, the conductance of the load, the conductance representing all losses in the input tank circuit and the magnitude of the negative conductance generated by the parametric operation. Evidently the gain tends to infinity as G approaches equality with G_{T1}.

The alternative way of using a circuit of the type illustrated in Fig. 5.1 is to take the output from the idler circuit, at a frequency other than the input signal frequency. In this case the load conductance does not appear in parallel with the input and is not added to the total loss conductance G_{T2} of the idler circuit when calculating the noise figure of the amplifier. In this case the noise figure is

$$F = 1 + \frac{G_1}{G_s} + \frac{G_{C2}}{G_{T2}}\frac{G}{G_s}\frac{f_1}{f_2} + \varphi_1(C) + \varphi_2'(C) \qquad (5.6)$$

where $\varphi_2'(C)$ differs slightly from $\varphi_2(C)$. The gain in this mode, with both circuits tuned to resonance, is

$$\text{Power gain} = \frac{4 f_2}{f_1}\frac{G_{L2}}{G_{T2}}\frac{G G_s}{(G_{T1} - G)^2} \qquad (5.7)$$

Note that in (5.6) and (5.7) the contribution to the conductance associated with the idler circuit from the capacitor, G_{C2}, and from the inductor, G_{L2}, is specified separately from the total conductance G_{T2}. This is necessary when the voltage applied across the resonant circuit contains components far from the resonance frequency and the overall balance of reactive components is critical. Since the gain is proportional to f_2/f_1 it may be advantageous to use this circuit as an up-converter, in contrast to the down-conversion of conventional superheterodyne operation; but this is not essential if $G_{T1} - G$ is made small enough.

Carter *et al.* (1969) quote a noise temperature of 130 K for an uncooled parametric amplifier and 23 ± 1 K for a parametric amplifier tunable from 3.7 to 4.2 GHz and helium cooled. Aitchison (1967) reported that a cavity parametric amplifier working at 4 GHz and cooled to 4 K had a noise temperature of 3 K.

The technique of cooled parametric amplifiers working in the frequency range 4 to 10 GHz became well established in connection with satellite communications, but the drive towards higher frequencies arose in three contexts: (1) high-definition radar, (2) radio astronomy and (3) a general desire to bridge the gap between radio and infra-red quasi-optical techniques. Daglish *et al.* (1968) pointed out that the possibility of raising the idler frequency in a parametric amplifier depends on the diode used as a variable capacitance in two ways: the diode must function satisfactorily as a variable capacitance at the higher pump frequency and the idler circuit resonant at frequency f_2 consists largely, if not entirely, of the stray capacitance and inductance of the diode. Schottky-type GaAs diodes are recommended and can be operated at 4 K. An example of development towards higher frequencies was described by Parrish and Chiao (1974). They replaced the varactor diode by a thin-film array of 80 Josephson-type junctions and had pump, idler and signal frequency arranged as $2f_p = f_i + f_s$. The junctions were of tin–tin oxide–tin and the predicted maximum usable frequency was that corresponding to the energy gap of tin, namely 273 GHz.

However, SIS mixers, super-Schottky diodes and cooled GaAs FETs are now used for the higher frequencies.

The conventional parametric amplifier employs a non-linear capacitor as active element but the SUPARAMP (Superconducting Unbiased PARametric AMPlifier) uses a series array of Josephson junctions as non-linear inductance for the active element (Feldman *et al.*, 1955), with the advantage the non-linearity is on a small scale and therefore the required pump power is low. Since the non-linearity of an unbiased Josephson junction is symmetrical about the zero of the pump cycle (unlike a varactor diode) it will produce no even harmonics and this places some constraints on the combination of pump, signal and idler frequencies which can be used. The usual arrangement is to have signal and idler frequencies equally spaced on either side of the pump frequency so that $2\omega_p = \omega_s + \omega_i$ and to use the SUPARAMP with a circulator as reflection amplifier of the signal. The Manley–Rowe relationships give

$$P_s/\omega_s = P_i/\omega_i \quad \text{and} \quad P_p/\omega_p + P_s/\omega_s + P_i/\omega_i = 0$$

Kadlec (1979) investigated the noise performance and reported an input noise temperature of 1 K for frequencies up to 10 GHz, with stable power gain of 20 to 30 dB.

5.4 The tunnel diode

Tunnel diodes can be used as negative-resistance amplifiers and have been so used in communication satellites: they are effective at low temperatures can can withstand thermal cycling. The passage of electrons from one energy band to another by tunneling is not subject to any form of 'thermalisation' or influence of space charge so it produces simple shot noise. There is also some Johnson noise in the series resistance of the semiconductor, but this is usually a small contribution and can be minimised by cooling. The noise figure of a tunnel diode therefore depends primarily on the ratio of the negative conductance (inverse of negative resistance) to average current at the working point. In order to obtain a homogeneous ratio of resistances, Nielsen (1960) replaced the shot noise in the diode, which is

$$\overline{i_d^2} = 2eI_0\,\Delta f$$

by thermal noise in a conductance $G_e = eI_0/2kT$. It is then possible to write the shot noise as though it were a thermal noise

$$\overline{i_d^2} = 4kTG_d(G_e/G_d)\,\Delta f$$

The stray resistance in the diode is physically a series element, so, in order to add its contribution to $\overline{i_d^2}$, so as to express the total noise at the terminals of the diode as a fluctuating current, it is necessary to represent the whole, including the stray shunt capacitance and series resistance of the diode, by a shunt equivalent circuit. G_d and G_s are then replaced by $G_{d'}$ and $G_{s'}$, the real parts of the corresponding admittances in the shunt equivalent circuit. (These admittances are complex and the real parts differ from the original conductances because of the stray capacitance and series resistance of the diode.) The result is a noise figure *for large gain* of

$$F = 1 + \frac{T}{T_0} \frac{\dfrac{G_e}{G_d} + \dfrac{G_{s''}}{G_{d''}}}{1 - \dfrac{G_{s''}}{G_d}} \tag{5.8}$$

where T_0 is a reference temperature. Since the relation of the shunt equivalent conductances to the basic parameters explicitly depends on the effect of diode capacitance C and series resistance R_s, one can define a cut-off frequency as

$$f_c = \frac{1}{2\pi C R_d} \sqrt{\frac{R_d}{R_s} - 1}$$

so that for $T = T_0$

$$F = \frac{1 + \dfrac{G_e}{G_d}}{\left(1 - \dfrac{R_s}{R_d}\right)\left(1 - \dfrac{f}{f_c}\right)^2} \tag{5.9}$$

For $f/f_c \ll 1$ and $R_s \ll R_d$ this becomes $F_0 = 1 + G_e/G_d$, depending solely on the properties of the material. For germanium, $F_0 = 3.1$ dB or $T_n/T_a = 1.04$.

The tunnel diode has also been used as an oscillator, with the merit that it can be incorporated in a superconducting cavity, the high Q of which is advantageous (Campisi and Hamilton, 1981).

5.5 SIS mixers

By resorting to superconductivity one eliminates resistance (as well as lowering temperature) so that the basis of the Nyquist theorem for thermal noise is destroyed, but the problem still remains of finding a device which gives good efficiency as a frequency-changer so that the signal is not lost in the noise of following amplifiers. To give good conversion efficiency (usually minimum conversion loss) a diode should appear non-linear on the scale of the voltage applied to it from signal plus local oscillator. The Superconductor–Insulator–Superconductor junction passes very little current at bias voltage below that corresponding to the energy gap (in electron-volts) of the superconductor, but at this point begins to pass a substantial current by tunneling so that there is a very sharp bend in the $I - V$ characteristic as shown in Fig. 5.3. One problem is to avoid the true Josephson effect, that application of steady bias produces r.f. current at a frequency given by the Josephson formula

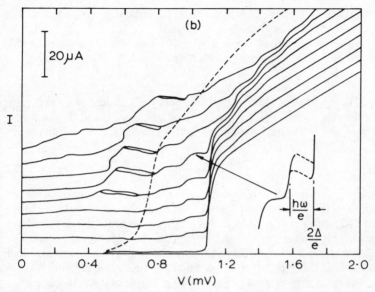

Fig. 5.3 Measured I–V curves for a tin junction at 1.5 K. Curves are for various local oscillator powers at 36 GHz, starting at about 0.2 nW and increasing upward in 2-dB steps. The inset shows in detail a region of negative resistance due to quasi-particle tunneling. (From McGrath et al., 1981)

$$v = 2eV/h \quad \text{or} \quad \omega = 2eV/\hbar$$

This can be done by applying a magnetic field parallel to the plane of the junction. However, the Josephson effect does not seem to be dominant at the higher voltages and currents, e.g. to the right of the dashed line in Fig. 5.3, and this constitutes the difference between the SIS junction and the Josephson junction. In terms of the quantum physics, the Josephson effect is believed to be due to tunneling by pairs of electrons (the pairs of the Cooper–Bardeen–Schrieffer theory of superconductivity) but there is also tunneling by single electrons which is known as quasi-particle tunneling. This occurs most readily when the valence band on one side of the barrier is level with the conduction band on the other side and is represented in Fig. 5.3 by the rapid rise of current at 1.1 mV. When voltage from a local oscillator is also applied, photon assisted tunneling produces steps in the characteristic below the critical voltage, as shown in Fig. 5.3, where the different curves correspond to different values of local oscillator power, and these steps can provide the acute non-linearity which is necessary for efficient mixing.

According to classical circuit theory, a resistive diode mixer must always show a conversion loss, i.e. the power at the new (i.f.) frequency must always be less than the input signal power. But Taur *et al.* (1974) showed that a mixer using a Josephson-type junction can have a conversion efficiency greater than unity and given by

$$\eta = C_{if}(R_{dyn}/R)\alpha^2 \tag{5.10}$$

where C_{if} is the output coupling efficiency, R_{dyn} and R are the inverse of the slope of the I–V curve at the operating point and the resistance shunted across the Josephson junction, and α is a factor given by

$$\alpha = \frac{\partial(I_0/I_c)}{\partial[(8P_{lo}/RI_c^2)^{1/2}]} \tag{5.11}$$

where I_0 is critical current in the presence of local oscillator power P_{lo} and I_c is the value of I_0 at $P_{lo} = 0$. Descriptively, α is the rate of change of I_0/I_c with the square root of eight times the ratio of local oscillator power to maximum d.c. dissipation in the Josephson shunt. The parameter α may vary between zero and infinity, but typical values quoted for a Nb–Nb point contact at 8 K are 0.23 and 0.18. A conversion gain of $\eta = 4$ is given as an example at 8 K. At 4.2 K some steps had negative slopes (compare the inset in Fig. 5.3) and operation at such a point led to relaxation oscillation with η as high as 50 provided that the i.f. was within a factor of 1.5 of the frequency of relaxation oscillation. Tucker (1979) has since deduced from quantum analysis (which is appropriate, since the behaviour of the SIS junction is explicable only by quantum theory) that conversion gain is possible.

McGrath *et al.* (1981) report a mixer input noise temperature of 9 ± 6 K for a tin–tin oxide–tin junction on a silicon substrate in a helium bath at 1.5 K and working at 36 GHz. Dolan *et al.* (1981) reported a diode noise temperature of 17 K at 115 GHz, with a conversion loss of 7.6 dB and a mixer noise temperature of 62 K. They used lead, containing In, Au and Bi, as junction material with lead oxide as the insulating layer. (One of the problems with Josephson and SIS junctions is to find materials and structure such that the junction will not be destroyed by thermomechanical stress during transitions between room and superconducting temperatures (Taur and Kerr, 1978).) The noise temperature rises very rapidly at higher frequencies. Sutton (1983) suggested the use of SIS mixers up to 400 GHz but reported a noise temperature of 305 K at 241 GHz.

5.6 Super-Schottky diodes

Tucker (1979) pointed out in a review paper on tunnel junction mixers that, if the semiconductor side of a Schottky diode is very highly doped, the depletion layer may be so thin that conduction is mainly by tunneling through the barrier rather than excitation over it; and cooling to superconducting temperature minimises both thermal excitation over the barrier and Johnson noise in the bulk resistance of the semiconductor. The scale of the non-linearity of such a diode (in electron-volts) is then kT or $\hbar\omega$, whichever is the greater; and $\hbar\omega \simeq kT$ at a frequency of 40 GHz and temperature of 1 K. The super-Schottky diode consists of a junction between a superconducting metal (usually lead) and a heavily doped semiconductor (usually GaAs); and the minimisation of junction capacitance, which is essential for use at very high frequencies, has led to the use of lead contacts of one to three micrometres in diameter. But, although the super-Schottky diode is a tunneling device and the metal contact is superconducting, the junction does not show the photon-assisted tunneling of the Josephson and SIS devices. It has the advantages of freedom from critical operating bias (needed to suppress Josephson effects and obtain optimum non-linearity in SIS devices) and greater thermomechanical robustness than some of the other types of junction. The *diode* temperature can be approximately the same as the bath temperature but there is considerable conversion loss so the *mixer* temperature is several times higher. Use of room-temperature Schottky diodes up to a frequency of 692 GHz (wavelength of 0.434 mm) has been envisaged by Fetterman *et al.* (1978) but with a noise temperature of around 7,000 K. Examples of the results which have been achieved at various frequencies and temperatures are shown in Table 5.1.

Table 5.1 NOISE TEMPERATURES ACHIEVED WITH SCHOTTKY AND SUPER-SCHOTTKY DIODES

Frequency (GHz)	Ambient (bath) temperature (K)	Diode noise temperature (K)	Mixer (M) or receiver (R) noise temperature (K)	Reference
9	1.06	≈ 5	—	Vernon *et al.*, 1977
30	1.2	1.2	6 (M)	McColl *et al.*, 1979
80	20	—	200 (R)*	Kollberg and Zirath, 1983
88	20	—	210 (R)	Vowinkel *et al.*, 1983
92	1	$\leqslant 3$	(22–25 dB)†	Dickman *et al.*, 1981
115	15	62	350 (R)	Dolan *et al.*, 1981
140	20	—	250 (R)	Vowinkel *et al.*, 1983
670–690	300	—	5,000–7,000 (M)	Fetterman *et al.*, 1978

* Predicted.
† Conversion loss.

5.7 Cooled FET amplifiers

Commercially available gallium arsenide FETs can be used at cryogenic temperatures with useful gains at frequencies up to 30 GHz. The use of GaAs at low temperature has already been assumed in the super-Schottky diode and the important point is that unlike silicon it does not suffer freezing out of carriers if suitably doped: one method is to dope at 10^{17} cm^{-3} with tin which has negligible excitation energy. This enables it to function as a FET with the high electron mobility providing an added advantage; and, in fact, the mobility increases at low temperature (due to the presence of polar scattering) so that the gain correspondingly increases. Leichti and Larrick (1976) found experimentally that most of the change occurs between 300 K and 200 K and the curve is flattening out at 90 K. For a particular FET (gate 1 μm long by 500 μm wide) working at 12 MHz they found a noise figure of 3.5 dB at 300 K but only 0.8 ± 0.5 at 90 K. The theory of noise in the GaAs FET has been examined by Podell (1981), who distinguished between two sources of noise: (1) Johnson noise in various extraneous resistances in the source-drain loop, which varies with frequency through the effect of stray capacitances associated with them; and (2) noise due to the passage of electrons through the FET channel, including hot-electron effects, which is not dependent on frequency. (See Chapter 4 on noise from hot electrons.) On overall temperature effect, Capello and Pierro (1984) estimated a gain improvement of 8 dB out of 25 dB when working a five-stage 22–24-GHz amplifier at 77 K instead of room temperature. At the lower frequency of 430 MHz and the lower temperature of 4 K, Prance *et al.* (1982) recorded a gain of 19 dB from a single stage with a noise temperature of about 10 K. Their application was to the output of a r.f. SQUID, the special feature being that both input and output impedance matching was achieved with the help of GaAs varactor diodes which were included in the low-temperature housing.

At a frequency where it can be used (up to 30 GHz) the FET amplifier has an obvious advantage over the Schottky diode mixer in that a diode mixer (with the possible exception of the SIS) has a conversion loss so that mixer noise temperature is higher than diode noise temperature and the noise temperature of the following amplifier is likely to be more important. A further advantage to be sought from the use of FET amplifiers is the possibility of obtaining broad band amplification from a multi-stage arrangement. Capello and Pierro (1984) reported a 10% bandwidth from a cascade of five single-ended stages giving a gain of 25 dB at 20–22 GHz when working at 77 K with a noise temperature of 170 K. The FETs used had a gate length of 0.5 μm: more recent FETs with gate length of

0.25 μm should give better results. Ayasli *et al.* (1984) built a multi-stage travelling-wave amplifier to give gain of 30 dB across the whole band 2–20 GHz, with a noise figure of 9 ± 1 dB. This appears to have been at room temperature, but it could presumably be cooled if desired.

A noise figure of around 1 dB (reported by Leichti and Larrick at 90 K) means that T_n is about $T/4$; and with further improvement in noise figure on cooling to 4 K the noise temperature may be down to tens of millikelvins, i.e. of order 0.01 K. (See Chapter 7, p. 154.)

5.8 The SQUID

'Superconducting Quantum Interference Device' is a precise theoretical description of the SQUID, though its practical form might be more readily recognised as 'Josephson junctions in an inductive loop', one form of which is illustrated in Fig. 5.4. Note that the Josephson junction, represented by a cross, is bi-directional and not a rectifying device. The Josephson effect was first observed* in junctions, typically using the three layers of metal–metal oxide–metal, as described in connection with SIS detectors. But the general term often used is 'weak link', i.e. the point of poor conductivity in the circuit where current passes by tunneling, which covers not only junctions constructed as three layers but also niobium point contacts and extremely narrow regions ('microbridges') in thin superconducting films. A junction is usually shunted by a normal (not superconduct-

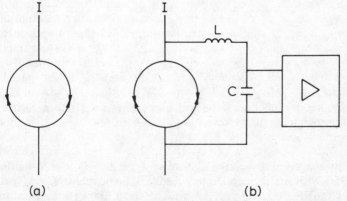

(a)   (b)

Fig. 5.4 (a) The d.c. SQUID, (b) coupling the d.c. SQUID by a circuit resonant at the 'carrier' frequency

* Josephson, 1965.

ing) resistance in order to prevent hysteresis in the I–V characteristic of the junction. The requirement is that the McCumber parameter β_c should be less than unity:

$$\beta_c = 2\pi r^2 i_c C/\varphi_0 < 1 \tag{5.12}$$

where i_c is the critical current of the junction, C is the junction capacitance and φ_0 the flux quantum which is approximately 2×10^{-7} cm^2-gauss (2×10^{-15} weber).

Figure 5.4, with two junctions in the loop, is for d.c. operation. In essence, the SQUID is a magnetometer, measuring the magnetic flux which passes through the loop. The combined current I flowing through the two junctions in parallel is controlled by the symmetry of the junctions, a symmetry which is disturbed by magnetic flux passing through the loop. The basic equation (Jaklevic *et al.*, 1964) is

$$I = I_0 \frac{|\sin \varphi_j e/\hbar|}{\varphi_j e/\hbar} \int \sin(\Delta y - \varphi_t e/\hbar) \tag{5.13}$$

where φ_j is flux which threads the current-carrying area of the junctions, φ_t is the flux threading the ring and Δy is the difference in phase of the wave function at the two junctions. The maximum current I_0 may be of the order of 1 μA to 1 mA, depending on the area of the junctions, and, since φ_j will normally be small compared with φ_t, due to the ratio of areas, the term of $(\sin x)/x$ form can be taken as constant and approximately unity while the second term varies cyclically. The latter goes through one cycle for a flux change through the loop of one flux quantum. (Note that as the current variation shown in formula (5.13) can be observed continuously within the cycle, the flux is not 'quantised' into discrete steps but is continuous on this scale.)

The noise energy per unit bandwidth of a d.c. SQUID is predicted to be

$$\varepsilon(1 \text{ Hz}) = 8kT/(R/L) \tag{5.14a}$$

where L is the inductance of the SQUID loop and R the shunt resistance of a junction. If the McCumber parameter of formula (5.12) is approximately unity, formula (5.14a) becomes

$$\varepsilon(1 \text{ Hz}) \simeq 8kT(LC)^{1/2} \tag{5.14b}$$

With a maximum voltage change of typically 60–100 μV the d.c. SQUID is usually operated in a feedback mode (phase locked loop) and the electronic circuit for this, using a 100-kHz carrier, was described by Clarke *et al.* (1976). An important feature is the use of a resonant circuit to match the low impedance of the SQUID to the high impedance of an FET amplifier, as shown in outline in Fig. 5.4(b).

The impedance transformation ratio depends on the Q of the resonant circuit LC and this may be contained with the SQUID in the low-temperature enclosure. In this type of circuit the noise will appear as modulation on a 100-kHz carrier and can be frequency-analysed.

The minimum possible value of noise in 1 Hz is the Planck quantum 6.6×10^{-34} joule-second or joule per hertz. Voss et al. (1980) reported a minimum noise of 17 h for niobium–niobium oxide–niobium, 5 h for lead alloy junctions and about 3 h for 'niobium nanobridge d.c. SQUIDs' in which the weak link in a ring of niobium film consisted of an area as small as 0.3 μm × 0.3 μm. The physical origin of the noise is assumed to be in the resistive shunt; and since this is at low temperature much of the literature speaks of 'zero-point fluctuation', i.e. the $\frac{1}{2}$ hv term in Equation (1.13). This is a concept which needs to be treated with care. The $\frac{1}{2}$ hv in (1.13) has also been described as 'vacuum fluctuation' which could not be exchanged and therefore could not be a source of noise. But according to (1.14) it is the value to which the average thermal energy tends as T tends to zero: it is not a separate entity, but is zero-point energy because it is the value which average thermal energy would take if T could be reduced to zero. (According to thermodynamics, zero temperature can never be attained.) The thermal noise is white and is dominant down to a certain frequency below which $1/f$ noise appears. Clarke et al. (1976) found this corner frequency to be 10^{-2} Hz; but the subsequent trend has been to reduce the area of junction, because this reduces capacitance and hence reduces thermal noise according to Equation (5.14b), and reduction in size in general increases $1/f$ noise. It is natural to apply to very small junctions the temperature-fluctuation theory of $1/f$ noise. Clarke and Hawkins (1976) calculated the $1/f$ noise to be expected from temperature fluctuations in various sizes of junction; but their formulae seriously under-estimated the $1/f$ noise in the 7-mm^2 junctions used by Clarke et al., which had a spectrum very close to f^{-1} below 10^{-2} Hz. Ketchen and Tsuei (1980) used 2.5-μm technology; and they found a corner frequency of about 1 kHz and low-frequency slopes of $f^{-0.9}$ at a temperature of 1.6 K and $f^{-0.75}$ at 4.2 K. The source of $1/f$ noise in SQUIDs is still as uncertain as the source of $1/f$ noise in general (Chapter 2).

There is also a radio-frequency form of the SQUID which uses only one junction. This depends on the $I–V$ characteristic of the junction (or weak link) being modified by magnetic flux through the loop so that a resonant circuit coupled to the SQUID can be modulated by the varying resistance introduced by the SQUID (see Fig. 5.5). The r.f. SQUID has been developed, particularly by the use of a cooled GaAs pre-amplifier which may be included in the cold chamber, to have a sensitivity comparable with that of the d.c. SQUID. It is, of course,

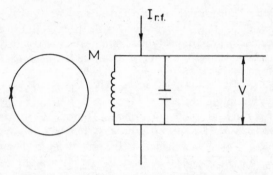

Fig. 5.5 The r.f. SQUID

free from $1/f$ noise; and the radio frequency is chosen as high as the availability of amplifiers allows. The predicted noise energy of the r.f. SQUID is

$$\frac{1}{K^2}\varepsilon(1\,\text{Hz}) \approx \frac{1}{K^2\omega_{\text{r.f.}}}(\pi a^2\varphi_0/2L + 2\alpha k T_{\text{eff}}) \qquad (5.15)$$

where K is the coupling coefficient between SQUID and r.f. coil (usually by mutual inductance), a and L are the area and inductance of the loop, α is a measure of the slope of the steps on the I–V curve and T_{eff} is the effective temperature of the amplifier. R.f. SQUIDs have been constructed with frequencies up to 500 MHz with cooled amplifiers and 10 GHz with room-temperature electronics. Their minimum noise has been of the order of 10^{-30} J/Hz.

REFERENCES

Aitchison, C. S. (1967). 'Low noise parametric amplifiers', *Philips Tech. Rev.*, **28**, 204–210

Ayasli, Y., Reynolds Jr., L. D., Vorhaus, J. L. and Hanes, L. K. (1984). '2–20 GHz GaAs travelling wave amplifier', *IEEE Trans.*, **MTT-32**, 71–77

Campisi, I. E. and Hamilton, W. O. (1981). 'A superconducting tunnel diode oscillator', *IEEE Trans.*, **MAG-17**, 838–840

Capello, A. and Pierro, J. (1984). 'A 22–24 GHz cryogenically cooled GaAs FET amplifier', *IEEE Trans.*, **MTT-32**, 226–230

Carter, J. W., Daglish, H. N. and Moore, P. (1969). 'A radiometer for measurement of the noise temperature of low-noise microwave amplifiers', *Radio Electron. Engr.*, **37**, 365–373

Clarke, J. (1980). 'Summary and conclusions', in *SQUID '80* (Ed. H. B. Hahlbohm and H. Lübbig), Walter de Gruyter & Co.; Berlin and New York, pp. 961–970

Clarke, J. and Hawkins, G. (1976). 'Flicker (1/f) noise in Josephson junctions', *Phys. Rev. B*, **14**, 2826–2831

Clarke, J., Goubau, W. M. and Ketchen, M. B. (1976). 'Tunnel junction d.c. SQUID: fabrication, operation, and performance', *J. Low Temp. Phys.*, **25**, 99–144

Daglish, H. N., Armstrong, J. G., Walling, J. C. and Foxell, C. A. P. (1968). *Low-noise Microwave Amplifiers*, Cambridge University Press; Cambridge

De Grasse, R. W. and Scovill, H. E. D. (1960). 'Noise temperature measurement on a travelling-wave maser preamplifier', *J. Appl. Phys.*, **31**, 443–444

Dickman, R. L., Wilson, W. J. and Berry, G. G. (1981). 'Super-Schottky mixer performance at 92 GHz', *IEEE Trans.*, **MTT-29**, 780–793

Ditchfield, C. R. (1968). 'Noise in masers and lasers', *Physical Aspects of Noise in Electronic Devices* (Conference at University of Nottingham, 11–13 September 1968), Peter Peregrinus; Stevenage, pp. 162–173

Dolan, G. J., Linke, R. A., Sollmer, T. C. L. G., Woody, D. P. and Phillips, T. G. (1981). 'Superconducting tunnel junctions as mixers at 115 GHz', *IEEE Trans.*, **MTT-29**, 87–91

Edrich, J., Harvey, P. C. and West, R. G. (1973). 'Low-loss 17°K coolable junction circulators for millimeter waves', *1973 European Microwave Conference*, vol. I, paper B9.5

Feldman, M. J., Parrish, P. T. and Chiao, R. Y. (1955). 'Parametric amplification by unbiased Josephson junctions', *J. Appl. Phys.*, **46**, 4031–4042

Fetterman, H. R., Tannenwald, P. E., Clifton, B. J., Parker, C. D., Fitzgerald, W. D. and Erickson, N. R. (1978). 'Far-IR heterodyne radiometer measurements with quasi-optical Schottky diode mixers', *Appl. Phys. Lett.*, **33**, 151–154

Heffner, H. and Wade, G. (1958). 'Gain, band width, and noise characteristics of the variable-parameter amplifier', *J. Appl. Phys.*, **29**, 1321–1331

Jaklevic, R. C., Lambe, J., Silver, A. H. and Mercereau, J. E. (1964). 'Quantum interference effects in Josephson tunneling', *Phys. Rev. Lett.*, **12**, 159–160

Josephson, B. D. (1965). 'Supercurrents through barriers', *Adv. Phys.*, **14**, 419–451

Kadlec, J. (1979). 'Noise performance and stability of a doubly degenerate unsaturated PARAMP', *J. Appl. Phys.*, **50**, 6443–6450

Ketchen, M. B. and Tsuei, C. C. (1980). 'Low frequency noise in small-area tunnel junction d.c. SQUIDs', in *SQUID '80* (Ed. H. B. Hahlbohm and H. Lübbig), Walter de Gruyter & Co.; Berlin and New York, pp. 227–235

Kollberg, E. L. and Zirath, H. H. G. (1983). 'A cryogenic millimetre-wave Schottky-diode mixer', *IEEE Trans.*, **MTT-31**, 230–235

Leichti, C. A. and Larrick, R. B. (1976). 'Performance of GaAs MESFETs at low temperatures', *IEEE Trans.*, **MTT-24**, 376–381

McColl, M., Bottjer, M. F., Chase, A. B., Pedersen, R. J., Silver, A. H. and Tucker, J. R. (1979). 'The super-Schottky diode at 30 GHz', *IEEE Trans.*, **MAG-15**, 468–470

McGrath, W. L., Richards, P. L., Smith, A. D., Butcher, R. A., Prober, D. E. and Santhaman, P. (1981). 'Large gain, negative resistance, and oscillations in superconducting quasi-particle heterodyne mixers', *Appl. Phys. Lett.*, **39**, 655–658

Manley, J. M. and Rowe, H. E. (1956). 'Some general properties of nonlinear elements— Part I, General energy relations', *Proc. IRE*, **44**, 904–913

Nielsen, E. G. (1960). 'Noise performance of tunnel diodes', *Proc. IRE*, **48**, 1903–1904

Parrish, R. J. and Chiao, R. Y. (1974). 'Amplification of microwaves by superconducting microbridges in a four-wave parametric mode', *Appl. Phys. Lett.*, **25**, 627–629

Podell, A. F. (1981). 'A functional GaAs FET noise model', *IEEE Trans.*, **ED-28**, 511–517

Pound, R. V. (1957). 'Spontaneous emission and the noise figure of maser amplifiers', *Ann. Phys.*, **1**, 24–32

Prance, R. J., Long, A. P., Clark, T. D. and Goodall, F. (1982). 'UHF ultra low noise cryogenic FET preamplifier', *J. Phys. E*, **15**, 101–104

Schulz-Dubois, E. O., Scovil, H. E. D. and De Grasse, R. W. (1959). 'Use of active material in three-level solid state masers', *Bell Syst. Tech. J.*, **38**, 335–352

Sutton, E. G. (1983). 'A superconducting tunnel junction receiver for 230 GHz', *IEEE Trans.*, **MTT-31**, 589–592

Taur, Y. and Kerr, A. R. (1978). 'Low-noise Josephson mixers at 115 GHz using recyclable point contact', *Appl. Phys. Lett.*, **32**, 775–777

Taur, Y., Claassen, J. H. and Richards, P. L. (1974). 'Conversion gain in a Josephson effect mixer', *Appl. Phys. Lett.*, **24**, 101–103

Tesche, C. D. (1981). 'A thermal activation model for noise in the d.c. SQUID', *J. Low Temp. Phys.*, **44**, 119–147

Tucker, J. R. (1979). 'Quantum limited detection in tunnel junction mixers', *IEEE J. Quantum Electron.*, **QE-15**, 1234–1258

Vernon, F. L., Millea, M. F., Bottjer, M. F., Silver, A. H., Pedersen, R. J. and McColl, M. (1977). 'The super-Schottky diode', *IEEE Trans.*, **MTT-25**, 286–294

Voss, R. F., Laibowitz, R. B., Ketchen, M. B. and Broers, A. N. (1980). 'Ultra low noise d.c. SQUIDs', in *SQUID '80* (Ed. H. B. Hahlbohm and H. Lübbig), Walter de Gruyter & Co.; Berlin and New York, pp. 365–380

Vowinkel, B., Gruener, K. and Reinert, W. (1983). 'Cryogenic all solid-state millimeter-wave receivers for airborne radiometry', *IEEE Trans.*, **MTT-31**, 996–1001

Chapter 6

Oscillator Noise

6.1 Introduction

At other than microwave frequencies one assumes that the significant impurity in the output waveform of an oscillator consists of harmonics of the desired frequency. Since the least frequency difference is then in 2:1 ratio, the harmonics are usually adequately frequency-filtered by the resonant circuit of the oscillator and by following circuits. In case this filtering is insufficient, for example due to a limited Q of the oscillator resonant circuit, it is possible to construct resonant circuits which offer a maximum impedance at frequency f but minimum at multiples of f (Groszkowski, 1933; see Fig. 12, p. 973). Figure 6.1 shows such a circuit which is parallel resonant at f but has a branch which is series resonant at $2f$. However, microwave oscillators tend to be perceptibly noisy, i.e. in addition to the desired frequency and its harmonics they generate appreciable 'noise sidebands' which represent both phase and amplitude modulation of the central or carrier frequency. The greater magnitude of this effect in microwave oscillators, as compared with oscillators working at lower frequencies, is associated with the higher noise measure of microwave active devices. The simplest approach is to assume that there is an independent white noise source which modulates the oscillator. Since the output of a white-noise source is random both in amplitude and in phase, it can be thought of as divided into two components (Mullen, 1960): one is of variable amplitude and produces AM, while the other is of variable phase and produces PM. A communication system will usually employ modulation of one type or the other, so that the receiver will largely ignore one of the types of noise. In considering the receiver response, it must also be remembered that the upper and lower sidebands of this noise are correlated, whereas white noise generated in the channel or in the receiver is uncorrelated across the whole band. The shape of the noise spectrum from the oscillator is also influenced by the characteristic of the resonator in the oscillator.

In theory, one ought to solve the non-linear differential equation of

117

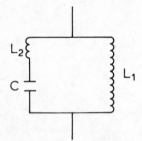

Fig. 6.1   Resonant circuit for the elimination of second harmonic. $(L_1 + L_2)C = 1/4\pi^2 f^2$; $L_2 C = 1/4\pi^2 (2f)^2$

the oscillator system in order to compute all the components of its output, after specifying all the sources of noise (Kutznetzov, 1955; Grivet and Blaquière, 1963). But, in practice, one can use the approximate method of a linear equation and parameters averaged over a cycle: this may be further justified because the shape of the characteristic of the non-linear active device may not be known with sufficient exactness to specify the coefficients in a non-linear differential equation.

Before proceeding to detail one must consider whether it is necessary to distinguish between *feedback* and *negative-resistance* oscillators which are shown schematically in Figs. 6.2(a) and 6.2(b). From the diagrams, it is clear that the function of the amplifier plus feedback in Fig. 6.2(a) is to produce the circuit effect of a negative resistance, so that for most purposes no distinction is necessary. (The feedback oscillator is more likely to have an asymmetric characteristic, and so produce even harmonics; but, as long as the procedure is by linearisation, the nature of the non-linearity does not matter.) It is assumed that the feedback impedance Z in Fig. 6.2(a) does not

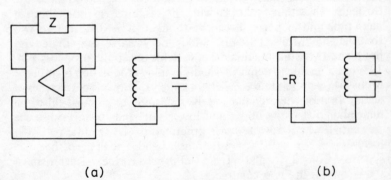

(a)   (b)

Fig. 6.2 (a) *Feedback oscillator*, (b) *negative-resistance oscillator*

contribute appreciable noise. In practice, a coupling is often incorporated in the resonant circuit.

The active element in an oscillator is usually run in a condition of 'hard limiting' so that the peak-to-peak amplitude of the (non-sinusoidal) oscillating voltage is approximately equal to the voltage available from the power supply. For small changes, the fundamental-frequency component of the oscillation is proportional to the supply voltage (this is the principle of the 'anode modulation' used in high-power radio telephony transmitters, though in this case over a wide range of voltage) so that the fractional modulation by noise of the amplitude is approximately equal to the ratio of the r.m.s. noise voltage to the supply voltage. On the basis of full shot noise in a mean current of 1 mA, the mean-square AM noise power per hertz would be over 150 dB below carrier power. The FM noise modulation, however, is much greater for low modulation frequencies.

The FM noise is closely related to the extent to which the oscillator frequency can be 'pulled' in synchronisation, and therefore an analysis of synchronisation may usefully accompany an analysis of oscillator noise.

The description of oscillator noise in terms of the noise measure of the active device employed makes it unnecessary to examine separately the noise performance of the various devices which may be used in oscillator circuits.

6.2 The analytical approach

Electronic oscillators (generators) normally rely on non-linearity of the active device to limit the amplitude of oscillation to a stable value, as a result of the average loop gain at fundamental frequency decreasing as amplitude increases. (An exception is the arrangement in which average loop gain is controlled in a feedback network, through a thermal or other device which is amplitude sensitive and has a time-constant long compared with the period of oscillation, so that the active device works in a linear regime. Such arrangements have been used for crystal-oscillator frequency standards (Meacham, 1938) and in Wien bridge oscillators in which a thermistor usually forms one arm of the bridge.) Both the amplitude stabilisation of simple oscillators and the synchronisation of coupled oscillators depend on non-linearity and it has long been customary to assume a power series representation of the non-linear characteristic of the form

$$i = a_0 + a_1 V + a_2 V^2 + a_3 V^3 \tag{6.1}$$

Higher powers have been neglected (from Van der Pol, 1934, to

Kurokawa, 1968). Groszkowski (1933) pointed out that the presence of harmonics in the oscillatory circuit slightly shifts the frequency at which inductive and capacitive energy storage balance, as they must do if the amplifying device which sustains oscillation is purely resistive; but this represents a liability to dependence of mean frequency on operating conditions, not immediately a source of noise. Chen *et al.* (1983) have pointed out that if the non-linearity contains a square law term any extraneous component of V at any frequency, such as low frequency noise, will modulate the oscillation; and this modulation noise can be suppressed by balancing the oscillator circuit so as to make it symmetrical, thus eliminating the even powers from the characteristic. They demonstrated the effect of balancing an oscillator working at 981.5 Hz and consisting of an operational amplifier with CR feedback. Balancing reduced the noise and eliminated the sidebands corresponding to a small marker voltage at 210 Hz which had been introduced. It should be noted, however, that this effect depends on the extraneous noise appearing as a *voltage*; and in most high-frequency oscillators the frequency-controlling circuit presents a low impedance at low frequency so that low frequency noise would not appear as a voltage which would then be modulated up to the oscillation frequency.

If the waveform is so 'hard limited' that the current approximates to a square wave, with amplitude of nth harmonic proportional to $1/n$, the 5th harmonic in the current will have an amplitude of 3/5th of the amplitude of the third harmonic. But, since the method of linearisation takes no account of the shape of the characteristic, the relative sizes of the various coefficients is immaterial. Some measure of the degree of saturation (limiting) is needed, and in fact this is apparent in Kurokawa's analysis and appears in a different form in Edson's, as discussed below.

The general method of finding the noise modulation is to add to the differential equation of an oscillator a term representing the output of a noise source (of unspecified physical origin) and assume that it acts as a small perturbation of the pure oscillation. One of the earliest analyses of this type was by Kuznetzov *et al.* (1955). They assumed that the disturbance was slow compared with the oscillator frequency, which corresponds to the more recent assumption that consideration is limited to frequencies close to the central oscillator frequency. Although the title of Kurokawa's paper (1968) refers to synchronised oscillators, there is a section of it which gives the noise in a free-running oscillator. The advantage of his approach is that it shows explicitly that the FM noise is related to the extent to which the frequency of oscillation can be modified by an injected voltage (sinusoidal or random) while the AM noise is related to the extent to which the oscillator is hard-limited in amplitude.

Kurokawa split the differential equation of a harmonic oscillator into two parts, representing amplitude and phase relations respectively, and added to each of them the output $e(t)$ of a noise source. This splitting of the oscillator equation can be justified symbolically in the $j\omega$ method of representing the steady state or by noting that sine and cosine components are orthogonal. This is only exact in a linear system since non-linearity produces harmonics which interact with the resonant circuit in such a way as to disturb the balance of 'imaginary' components; but this effect is very small and can be ignored in noise calculations.* Starting with the oscillator equation

$$L\frac{\mathrm{d}i}{\mathrm{d}t} + (R_0 + R_i + \bar{R})i + \frac{1}{C}\int i\,\mathrm{d}t = e(t) \qquad (6.2)$$

where R_0 is the external load resistance, R_i the resistance of the oscillatory circuit and $-\bar{R}$ the average over an oscillation cycle of the negative resistance generated by the oscillator, all expressed as a series combination with L and C. Kurokawa assumed a basic steady state solution

$$i(t) = A_1(t)\cos(\omega_t + \varphi_1) + A_2(t)\cos(2\omega t + \varphi_2) + \cdots$$

with noise as a small perturbation. This solution is substituted into (6.2) and multiplication by $\cos(\omega t + \varphi)$ and $\sin(\omega t + \varphi)$ respectively and integration over the period τ_0 of ω results in the two equations

$$(-\omega L + 1/\omega C) - (L + 1/\omega^2 C)\frac{\mathrm{d}\varphi}{\mathrm{d}t} = \frac{2}{A\tau_0}\int_{t-\tau_0}^{t} e(t)\sin(\omega t + \varphi)\,\mathrm{d}t \qquad (6.3)$$

$$(L + 1/\omega^2 C)\frac{\mathrm{d}A}{\mathrm{d}t} + (R_i + R_0 - \bar{R})A = \frac{2}{A\tau_0}\int_{t-\tau_0}^{t} e(t)\cos(\omega t + \varphi)\,\mathrm{d}t \qquad (6.4)$$

(The integral of $\sin^2 \omega t$ or $\cos^2 \omega t$ over a period τ_0 is $\tau_0/2$; and in (6.3) and (6.4) the factor $\tau_0/2$ is transferred to the right-hand side as $2/\tau_0$.) The notation is that R_0 and R_i are the load and internal (loss) resistances, $-\bar{R}$ the average over the cycle of the negative resistance, A the amplitude of oscillation (with equilibrium value A_0) and $e(t)$ the noise voltage. Two further steps are necessary to obtain a result which is compatible with engineering practice. The first is to transform $e(t)$, a time function which in Fourier terms contains all frequencies, into a spectral density, so that the noise in a limited frequency band can be calculated. Kurokawa does this by assuming that $e(t)$ is made up of a

---

* For the small effect on oscillator frequency, see Groszkowski, 1933.

number of narrow pulses, and that even after integrating over the period of a cycle of oscillation the noise can still be considered as made up of pulses which are short compared with τ_0. (Yet he assumed that $e(t)$ did not change appreciably within τ_0.) One then deduces a flat spectrum, as in the case of shot noise (Appendix I). The second step is to relate the magnitude of noise to known circuit parameters. Without specifying the mechanism of the noise, one can use the noise measure M of the active device, with the definition

$$\overline{e_B^2} = M 4kT|R|B \tag{6.5}$$

where $\overline{e_B^2}$ is the mean-square noise voltage within a frequency band B. (The band B is usually taken to be narrow, e.g. when written df, the essential requirement being that $\overline{e^2}$ should be constant over B.) If $1/f$ noise is also present it is represented as a multiple of thermal noise, so that the total noise is

$$e_B^2 = M(1 + f_0/f) 4kT|R|B \tag{6.6}$$

where f_0 is the corner frequency for the combination of $1/f$ noise with white noise. A weakness of this approach is that it does not consider the possibility of specific phase effects, such as fluctuation of the input capacitance of the active device, in addition to the phase-modulating effect of gaussian amplitude fluctuations.

6.3 FM and AM noise

The form of the expressions for FM and AM noise, being functions of the susceptibility of the oscillator to modulation in frequency and amplitude, depend on the way in which one represents the amplitude limiting and frequency pulling characteristics of the oscillator. Kurokawa represents the amplitude limiting in terms of the change of average (negative) resistance with amplitude: plot a graph of $\bar{R}$ versus A, take a tangent to this curve at the working point where $A = A_0$ and denote by $R_0 + R_i + KA_0$ the value of $\bar{R}$ at which this tangent intercepts the $\bar{R}$ axis, as shown in Fig. 6.3. (At the working point, $\bar{R} = R_0 + R_i$.) Stability requires that K be positive, i.e. that the negative resistance at the working point decreases as A increases. The range of frequency pulling, $\Delta\omega_0$, by an injected sinusoidal voltage oscillation of amplitude a_0 is

$$|\Delta\omega_0| \leqslant (a_0/A_0)(1/2L) \tag{6.7}$$

Note that since this result has been obtained from a linearised equation, substituting $a_0 \cos(\omega_s t + \varphi_s)$ for the $e(t)$ in Equation (6.2), it includes no explicit reference to non-linearity. Neglecting terms with

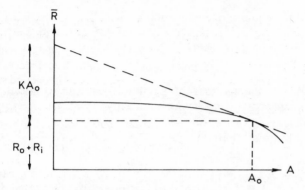

Fig. 6.3 Kurokawa's K measure of saturation or limiting

$(\omega + \omega_s)^2$ in the denominator, compared with those with $(\omega - \omega_s)^2$, Kurokawa's expressions for the intensity of noise modulation, measured in amperes² per hertz, are as follows:

$$\text{FM noise*} = \frac{|e|^2}{8L^2} \left[\frac{1}{(\omega - \omega_s)^2 + (|e|^2/4L^2 A_0^2)^2} \right] \qquad (6.8a)$$

$$\text{AM noise} = \frac{|e|^2}{8L^2} \left[\frac{1}{(\omega - \omega_s)^2 + (KA_0/2L)^2} \right] \qquad (6.9a)$$

Note that the expression for AM noise includes the 'saturation' coefficient K while that for FM includes the synchronising factor $2LA_0$ of Equation (6.7).

Equations (6.8a) and (6.9a) refer to the noise *current* in the equivalent series circuit. Kurokawa also quotes the output *voltage* spectrum in terms of available power, $P_a = 2Ri^2$, which implies that the oscillator is loaded for maximum output power. (The noise performance would be improved by light loading of the oscillator.) The result for this matched load, with noise power N and oscillator power P_0 is

$$P(f) = \frac{\omega_0^2}{2Q_{ext}^2} N \left\{ \frac{1}{(\omega - \omega_0)^2 + (\omega_0^2/4Q_{ext}^2)^2(N/P_0)^2} \right.$$

$$\left. + \frac{1}{((\omega - \omega_0)^2 + (\omega_0/Q_{ext})^2(KA_0/2R_0)^2} \right\}$$

$$= AM \text{ noise} + FM \text{ noise} \qquad (6.10)$$

* $|e|^2$ is to be taken as a spectral intensity, whereas $e(t)$ in Equation (6.2) was a 'waveform' containing spectral components of all frequencies.

Edson (1960) used different notation with $1/G$ in place of Kurokawa's $R_i + R_0$, $|e|^2$ replaced explicitly by $2kT$ and saturation represented by a parameter S defined by

$$SG = I/\Delta V \tag{6.11}$$

where ΔV is the increase of amplitude caused by injection of synchronizing current I. The noise intensity in volt^2 per hertz is then

$$\text{FM noise: } V_1^2(\omega) = \frac{2kTG}{(\omega_0 CkT/2PQ^2)^2 + 4C^2(\omega - \omega_0)^2} \tag{6.8b}$$

$$\text{AM noise: } V_2^2(\omega) = \frac{2kTG}{S^2G^2 + 4C^2(\omega - \omega_0)^2} \tag{6.9b}$$

Note that, as expected, the saturation parameter S appears in the formula for AM noise. In both equations the right-hand side has the dimensions of admittance times $2kT$.

By dropping the term in $|e|^2$ in the denominator of (6.8a), because it is small compared with $\omega - \omega_0$ for all but the smallest values of the latter, Hamilton (1978) was able to put the noise-to-carrier power ratios for bandwidth B in simpler form:

$$\text{FM noise: } \frac{N}{C} = \frac{1}{2} \frac{MkTB}{P_0} \left(\frac{f_0}{Q_{ext} f_m} \right)^2 \tag{6.8c}$$

$$\text{AM noise: } \frac{N}{C} = \frac{1}{2} \frac{MkTB}{P_0} \frac{1}{(S/2)^2 + (Q_{ext} f_m/f_0)^2} \tag{6.9c}$$

where $Q_{ext} = \omega_0 L/(R_0 + R_i)$, $f_m = \omega - \omega_0$ and $(S/2)^2$ replaces the term $(KA_0/2L)^2$.

6.4 Other sources of noise

The mechanism of the noise depends upon the type and characteristics of active device used but its effect depends also on the circuit parameters. Fluctuations in current may cause fluctuations in input capacitance, e.g. at the base in a junction transistor or the gate in a FET, so that a resonant circuit with a large ratio of capacitance to inductance may be helpful. In addition, it has been shown by Chen et al. (1983) that a square law term in the i/v characteristic of the maintaining device will cause low-frequency noise voltage to modulate the oscillation so as to produce noise sidebands which are additional to those produced by components of device noise at the oscillator frequency. They demonstrated this effect with an oscillator which had a frequency of 981.5 Hz and was based on an operational amplifier to provide the gain. A signal of 10 mV at 210 Hz was injected in order to identify the occurrence of modulation. There was provision to adjust the oscillator system to exact symmetry, so as to eliminate the even powers from the i/v characteristic, which caused the 210-Hz sidebands to disappear from the output and the noise sidebands at lower frequencies to be reduced.

In other cases the noise associated with avalanche multiplication or with fluctuation in the timing of creation of domains in a Gunn device may predominate over the basic thermal noise, but will be included in a noise measure M. But in all cases the noise modulation varies inversely with the working Q of the resonant circuit ($1/Q^2$ in Equation (6.8c)). Frequency modulation may also result from changes in phase within the oscillator circuit, e.g. due to fluctuations in transit time or to phase jitter in a frequency-multiplication chain, and the change in frequency necessary to cause a compensating change in phase of a simple resonant circuit varies inversely with Q. There is a limit to increasing Q, since the oscillator must follow desired modulation, but this limit is not usually reached in microwave oscillators. From (6.9a) it can be seen that the corner frequency, beyond which AM noise ceases to be constant, is given by

$$\omega - \omega_s = f_c = K A_0/2L \tag{6.12}$$

Now KA_0 is a resistance of the same order as the circuit resistance, Fig. 6.3, so it follows that

$$K A_0/2L \approx R/2L = \omega R/2\omega L = \omega/2Q \tag{6.13}$$

This approximately represents the highest frequency to which the resonant circuit (without the active device) could respond. The exact expression $KA_0/2L$ represents the highest modulation frequency to which the complete oscillator can respond. Hence the AM noise can be considered constant within the frequency band of interest in a practical microwave oscillator. Another way of making the phase change rapidly with frequency is to include a delay line in the feedback circuit, instead of relying on high Q, and the application of this to tunable oscillators for superheterodyne receivers has been discussed by Underhill (1978).

6.5 Practical results

Equation (6.9c) shows AM noise constant as long as $S/2 \gg (Q_{ext} f_m/f_0)^2$ and FM noise falls as the square of f_m, the sideband frequency, if the internal noise source produces white noise. The addition of $1/f$ noise produces a combined $(FM + AM + 1/f)$ noise spectral density which falls initially as $1/f_m^3$, then as $1/f_m^2$ for FM noise above the corner frequency of the $1/f$ noise, thereafter tending to a constant value representing the AM noise plus any added white noise from other parts of the system as far as the edge of the working bandwidth. If the desired frequency is obtained by multiplication of the frequency of a more stable oscillator at lower frequency, one has to allow also for (1) the fact that any phase noise in the basic oscillator is correspondingly multiplied and (2) that there may be phase jitter in the multiplication chain. Hamilton (1978) compared three methods of obtaining an output of 10 GHz: (1) a 100-MHz 5th overtone crystal, followed by frequency multiplication of 100 times, without jitter; (2) a 1-GHz cavity oscillator, frequency multiplied by 10; and (3) a Gunn oscillator working directly at 10 GHz. His results were summarised in the form reproduced in Fig. 6.4, from which it can be seen that the Gunn oscillator is best in respect of noise modulation at frequencies above 1 MHz, but the crystal oscillator times 100 is best below 100 kHz. The noise to carrier ratio, expressed as the ratio of noise power in 1-Hz bandwidth to carrier power, ranges for different oscillator arrangements between slightly positive to minus 39 dB at 10 Hz and -120 to -160 dB at high modulation frequencies. Note that, although the difference between best and worst oscillator is about 40 dB at both extremes of sideband frequency, the best/worst oscillator types are interchanged between the two extremes.

Much better carrier/noise ratio can be obtained in a sufficiently narrow bandwidth by using a high Q resonator, which implies light loading of the oscillator. For example, the tunnel diode of Campisi

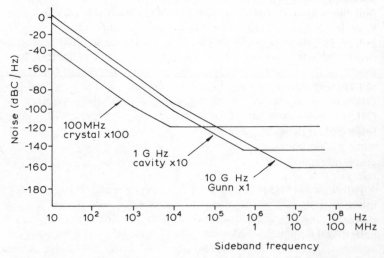

Fig. 6.4 *Relative noise of different types of 10 GHz oscillator. (After Hamilton, 1978)*

and Hamilton with a superconducting cavity or cavities (Chapter 5) had a noise/carrier power ratio in the range 0.1 to 1 Hz from the carrier of −210 dB with a single cavity or −240 dB with three cavities.

According to Hamilton, the Gunn diode has a noise measure of some 24 dB for frequencies above a few kHz, while a silicon avalanche diode has a noise measure above 100 Hz of about 35 dB. (The frequencies quoted are the approximate corner frequencies for $1/f$ noise.) Silicon avalanche devices include Impatt, Trapatt and Barrit diodes; and the noise properties of avalanche multiplication are described in Chapter 4.

The possible seriousness of the problem of oscillator noise is shown by Besada (1979), who quoted an example of a 100-GHz oscillator having only 16% of its power in the carrier, the rest being in noise.

6.6 Noise reduction by synchronisation

An obvious suggestion is that the purity of frequency of output of a microwave oscillator might be improved by synchronising it with a comparatively noise-free oscillator of lower power. Synchronisation depends on the non-linearity of the oscillator, which is usually taken only as far as a cubic term. The conditions were definitively established by Van der Pol (1934)* and re-examined much later in connection with the synchrodyne communication system, which requires a local oscillator in the receiver to be synchronised with the in-coming carrier (Tucker and Jamieson, 1956). The first problem is to find the noise-free reference oscillator. One thinks of oscillators of lower frequency, e.g. a crystal oscillator, as much better behaved; and, indeed, Hamilton's data (Fig. 6.4) suggests that at the lower modulation frequencies, where $1/f$ noise predominates in microwave devices, the 10-GHz output obtained from a 100-MHz crystal oscillator plus × 100 frequency multiplication is up to 40 dB better than either a 1-GHz cavity oscillator with × 10 frequency multiplication or a 10-GHz Gunn diode. One assumes that the stable oscillator will be of lower power than the one which is to be controlled by it, so that the work of Tucker and Jamieson (1956) on suppression of noise or amplitude modulation on the *weaker* of two oscillations is not relevant.

Kurokawa (1968) investigated synchronisation by the same method as he used for noise, substituting an injected synchronising current $a_0 \cos \omega_s t$ for the noise $e(t)$ in Equation (6.2). With this, one finds that the right-hand side of (6.3) evaluates to $(a_0/A) \sin \varphi$ where φ

---

* Van der Pol's equation is well known in mathematics.

is the phase angle between components at angular frequencies ω and ω_s. φ must be constant for synchronisation and $\sin \varphi < 1$ which leads to

$$|\Delta \omega| \leqslant a_0/2LA \tag{6.14}$$

where $\Delta \omega = \omega - \omega_s$. If $a_0 \ll A_0$, then $A \simeq A_0$ in Equation (6.14). But synchronisation always leads to an *increase* in amplitude of the stable oscillation, given by

$$\Delta A_0 = (a_0 \cos \varphi)/K A_0 \tag{6.15}$$

Since $\varphi = 0$ when $\omega_s = \omega$, the maximum magnitude of the change im amplitude is

$$\Delta A_{max} = (a_0/A_0)/K \tag{6.16}$$

i.e. it depends on the ratio of component amplitudes and on the degree of saturation as measured by K. Kurokawa was investigating the effect on an originally noisy oscillator of synchronising it by means of a current of smaller amplitude from a noise-free oscillator. It is not usually possible to provide a noise-free oscillator but, for the sake of completeness, his results are summarised as follows. The FM noise within the bandwidth for which $\omega - \omega_s$ is small is reduced by replacing $(|e|^2/4L^2A^2)^2$ in (6.8) by $a_0^2 \cos^2 \varphi_0/4L^2A_0^2$ where a_0 is the amplitude of synchronising current and φ_0 is the phase angle between synchronising and synchronised oscillation: φ_0 is of course zero if the free-running frequency of the controlled oscillator coincides with that of the synchronising oscillator. The replacement of noise intensity by synchronising intensity (squared amplitude) increases the denominator of (6.8a) and so reduces the FM noise. The AM noise is multiplied by a factor

$$\frac{(\omega - \omega_0)^2 + a_0^2/4L^2A_0^2}{(\omega - \omega_0)^2 + a^2 \cos^2 \varphi_0/4L^2A_0^2} \tag{6.17}$$

This is unity when $\varphi_0 = 0$ but is > 1 for the rest of the possible range of $-\pi/2 < \varphi_0 < \pi/2$. In general, therefore, AM noise would be *increased* as a result of synchronising with a noise-free oscillator. Kurokawa considered the case of synchronising by a noisy oscillator to be too complicated for analysis to be of practical use; but one can see qualitatively the results to be expected. Since the phase of the oscillation depends jointly on both oscillators, synchronising by a noisy oscillator must increase the FM noise, but probably with the contribution from the synchronising oscillator restricted by the ratio a_0^2/A_0^2 and by $\cos^2 \varphi$. The contribution to AM noise power will include an increase by the factor (6.17) but will be restricted by being proportional to the square of ΔA_0 as given by (6.15).

An interesting possibility is that if n similar oscillators are all mutually synchronised, the noise-to-carrier power ratio of the combination should be $1/n$th of that of one of the oscillators alone. According to the abstract of a Japanese paper (Okabe, 1977), this is theoretically possible for any number of oscillators and has been verified experimentally for two Impatt oscillators at X-band frequency. This seems right for FM noise, but presumably it would not apply also to AM noise.

REFERENCES

Besada, J. L. (1979). 'Influence of local oscillator phase noise on the resolution of millimetre-wave spectral-line radiometers', *IEEE Trans. Inst. Meas.*, **IM-28**, 169–171

Chen, H. B., van der Ziel, A. and Amberiadis, K. (1983). 'Reduction of the low frequency sidebands in oscillators', in *Noise in Physical Systems and $1/f$ Noise* (Ed. M. Savelli, G. Lecoy and J-P. Nougier), North-Holland Physics Publishers; Amsterdam, pp. 333–335

Edson, W. A. (1960). 'Noise in oscillators', *Proc. IEEE*, **48**, 1454–1466

Grivet, P. and Blaquiere, A. (1963). 'Non-linear effects of noise in electronic clocks', *Proc. IEEE*, **51**, 1606–1614

Groszkowski, J. (1933). 'The interdependence of frequency variation and harmonic content and the problem of constant frequency oscillation', *Proc. IRE*, **21**, 958–981

Hamilton, S. (1978). 'FM and AM noise in microwave oscillators', *Microwave J.*, **21**(6), 105–109

Kurokawa, K. (1968). 'Noise in synchronised oscillators', *IEEE Trans. Microwave Theory Tech.*, **MTT-16**, 234–240

Kurokawa, K. (1969). 'Some basic characteristics of broadband negative resistance oscillator circuits', *Bell Syst. Tech. J.*, **48**, 1937–1955

Kuznetsov, P. I., Stratonovich, R. L. and Tikhonov, V. I. (1955). 'The effect of electrical fluctuations on a vacuum tube oscillator', *Sov. Phys.—JEPT*, **1**, 510–514

Meacham, L. A. (1938). 'The bridge stabilised oscillator', *Proc. IRE*, **26**, 1278–1294

Mullen, J. A. (1960). 'Background noise in non-linear oscillators', *Proc. IRE*, **48**, 1467–1473

Okabe, Y. (1977). 'Noise in multiple microwave oscillator systems' (in Japanese), *J. Fac. Univ. Tokyo, Ser. B*, **34**, 71–87 (English abstract: *Sci. Abs. B*, 1978–5480)

Tucker, D. G. and Jamieson, G. G. (1956). 'Discrimination of a synchronised oscillator against interfering tones and noise', *Proc. IEE, Part C*, **103**, 129–138

Underhill, M. J. (1978). 'Comparison of the noise performance of some oscillators for tunable receivers', *IERE Conference on Radio Receivers and Associated Systems, Southampton, July 1978*, pp. 237–252

Van der Pol, B. (1934). 'Non-linear theory of electric oscillators', *Proc. IRE*, **22**, 1051–1086

Chapter 7

Noise in Radiation Detectors

7.1 Introduction

This chapter falls into four sections:

(1) The detection of optical signals including photon counting and hence photon-correlation spectroscopy ('the Malvern correlator').
(2) The detection of signals in optical-fibre communication systems.
(3) Detectors for the medium and far infra-red, both photo-electric detectors and thermal detectors in which the incident radiation is primarily converted to heat.
(4) The attempted detection of the gravitational waves which are predicted by Einstein's General Theory of Relativity (and sometimes in slightly different form, by other post-Newtonian systems of mechanics).

The detection of weak magnetic fields is dealt with in Chapters 5 (the SQUID) and 3 (the fluxgate magnetometer).

7.2 Optical signals

Since the invention of the laser it has become important to distinguish between *thermal* sources of radiation (incandescent lamps, stars, etc.) and *coherent* sources (mainly lasers), because the former are non-degenerate and the latter degenerate in the terms used in quantum theory. As in other applications of quantum theory, degeneracy depends upon the number of similar particles (photons of the same polarisation) in a unit cell in phase space, the latter being in this case the 'coherence volume'. If the unpolarised light from a thermal source is regarded as being equally divided between two perpendicular directions of polarisation, the degeneracy parameter is (Mandel and Wolf, 1965)

$$\delta \approx \tfrac{1}{2}(c^2/v^2)E_v \tag{7.1}$$

where v is the frequency and E_v the number of photons corresponding to this frequency which are emitted per unit area, unit time and unit frequency interval into unit solid angle. Equation (7.1) is clearly equivalent to $\delta \approx \frac{1}{2}\lambda_v^2 E_v$ where λ_v is the wavelength corresponding to frequency v. The squared length is cancelled out by the 'per unit area' factor in the specification of E_v, and the other factors are together dimensionless, so δ is purely numeric. For black-body thermal sources it can also be expressed as

$$\delta = [\exp(hv/kT) - 1]^{-1} \qquad (7.2)$$

For v in the middle of the visible region of the spectrum, $\delta \approx 1$ for $T = 3 \times 10^4$ K, but for thermal sources δ is usually of the order 10^{-3}. Lasers, on the other hand, have a high surface brightness concentrated in a very narrow frequency band and so a high value of degeneracy such as $\delta = 10^{14}$.

Weak light may be applied to a photomultiplier, the internal gain of which is sufficient to produce a measurable output current when a few electrons are emitted from the cathode. At light intensities too high for quantum fluctuations to be significant the photomultiplier has both the shot noise inherent in the primary photo-current and noise due to the randomness of the secondary-emission multiplication process at each stage of the multiplier. If the mean number of incident electrons per unit time is n_0 the variance (mean-square variation) of the number of incident electrons is also n_0. With a mean amplification A by secondary emission, the mean-square fluctuation in the amplified current is then $A^2 n_0$. The amplification factor A is also subject to random fluctuation, so that whereas the mean output for a fixed input n_0 is $M = An_0$, there will also be a fluctuation $\overline{\delta M^2}$ which for a Poisson distribution would have the value M. In fact, the variation of amplification about the mean need not necessarily be that indicated by the Poisson law, so it will be written $\overline{\delta M^2} = \alpha M = \alpha An_0$. The total fluctuation after one stage of amplification is then

$$\langle (M - \bar{M})^2 \rangle_{ave} = A^2 n_0 + \alpha A n_0 \qquad (7.3)$$

By taking the variance given by (7.3) as the fluctuation input to the second stage, multiplying by A^2, and adding the variance arising from the fluctuations in the A of the second stage, then repeating for subsequent stages, one finds the variance of the output after m stages; and the variance in number of electrons can be converted to a variance in current within a narrow bandwidth in the same way as for shot noise in a diode, leading to the formula for output noise current

$$\overline{\Delta I_{dv}^2} = 2ei_0 A^{2n} \left[1 + \alpha\left(\frac{1}{A} + \frac{1}{A^2} + \cdots \frac{1}{A^n}\right)\right]$$

$$= 2ei_0 A^{2n} \left[1 + \frac{\alpha[1 - (1/A^n)]}{A - 1}\right] dv \tag{7.4}$$

In the special case of $\alpha = 1$ this reduces to Zworykin's formula (Zworykin and Ramberg, 1949)

$$\overline{\Delta i_{dv}^2} = 2ei_0 \, dv \, \frac{A^{2n+1} - A^n}{A - 1} \tag{7.5}$$

$$\simeq 2ei_0 \, dv A^{2n+1}/(A - 1) \tag{7.6}$$

since A^n is usually large, say 10^6. The signal power is increased by a factor A^{2n}, so that the output signal/noise ratio is worse than that at the input by a factor which according to (7.6) is approximately $A/(A-1)$. With a typical amplification ratio of 10^6 in 10 stages, A is roughly 4 so that the signal/noise ratio is deteriorated by a factor of 4/3.

A more serious problem in practice is that of 'dark current', i.e. cathode emission in the absence of light, which is presumed to be largely thermionic and arising both from the photocathode and from the multiplier electrodes. If i_d is the dark current and i_s the mean signal current, the initial signal/noise ratio (in power) is

$$\left[\frac{S}{N}\right]_{cathode} = \frac{i_s^2}{2e \, dv(i_s + i_d)} \tag{7.7}$$

Note that, in principle, any value of i_s, however small, could be detected by narrowing the bandwidth dv; for very small bandwidths this operation would be described as lengthening the integrating time of the circuit. A typical noise specification is that of the R.C.A. multiplier phototube type 1 P21. With the light source (tungsten at 2,870°K) interrupted at 90 Hz and using an amplifier bandwidth of 1 Hz, the condition of unity signal/noise ratio is reached at a light intensity of approximately 4.5×10^{-13} lumens if the temperature is 20°C. Refrigerating the photocell lowers the noise level by about one decade of equivalent light intensity for 80°C drop in temperature.

It may be noted that according to Schwantes et al. (1956) no flicker noise is found in photomultiplier tubes, though there is flicker noise in thermionic valves which use secondary-emission amplification: the latter presumably arises from the thermionic cathode. Chenette et al. (1957) have therefore proposed the use of a photomultiplier as standard noise source for frequencies of 1 Hz upwards, but photo-multipliers suffer from a fatigue effect in that gain drifts for a considerable time after a change in working level. The noise output of a photomultiplier is considerable, being equivalent to a noise-diode current

$$i_{eq} = HA^2 i_p \qquad (9.8)$$

where i_p is the primary photocurrent, A the current amplification ratio and H a factor of the order of 1.5 to allow for statistical fluctuation in the current amplification.

If the light is weak enough for quantum fluctuations to be significant, it is usually assumed that the photons arrive independently at random; i.e. the numbers of photons received in successive short time intervals have a Poisson distribution. If the light can be applied to a photomultiplier of which the gain is sufficient to produce a detectable output current when a single electron is emitted from the cathode, one has the possibility of 'photon counting', though more precisely it is photo-electrons which one hopes to count. The quantum efficiency (ratio of number of photoelectrons to number of photons) is less than unity and the fluctuation in numbers of photoelectrons is due both to fluctuations in number of photons and to fluctuations in quantum efficiency. The variance in number of photons is given by

$$\langle (\Delta n)^2 \rangle = \langle n \rangle + \beta^2 \langle (\Delta U)^2 \rangle \qquad (7.9)$$

where ΔU is the fluctuation in the light wave field. For laser light ΔU is negligible, but for thermal radiation $\langle (\Delta n)^2 \rangle$ is greater than $\langle n \rangle$ due to the field fluctuations (Bertolotti in Cummins and Pike, 1974). The chief difficulty is the 'dark current'. Apart from leakage, which should be eliminated, there is the problem of electrons which are spontaneously emitted from the cathode, mainly as a result of thermal excitation, though a few may be due to cosmic rays. The thermal emission can be greatly reduced by cooling, since the excitation energy includes a factor $\exp(-e\varphi/kT)$ and division both by time and by area is used to make the number of spontaneous emissions in any sub-division small compared with the number of photoelectrons which it is desired to count.

7.3 Photon counting

The combination of photon counting with temporal correlation leads to spectroscopy with extraordinarily high resolution (Cummins and Pike, 1974). If a spectral line has a width δv on a centre frequency v, the classical method of interferometry by difference of path length requires a length of $c/\delta v$ to achieve a difference of one wavelength between extreme components. If δv is 100 MHz, this indicates a path length of 3 m, which is on the limit of practicability. Alternatively, there must be a corresponding amplitude variation at 100 MHz which can be detected in the amplitude trace which is produced by counting the numbers of photons in successive short intervals, e.g. intervals of one nanosecond. The frequency analysis of the amplitude variation can then be obtained from the autocorrelation function of the time series formed by the succession of photon counts. The number of photons in each count may be small enough for statistical fluctuations to be important. The randomness in the photoelectric detector follows the Poisson law, but the photons may form a degenerate system having an Einstein–Bose distribution with a different value of variance: detailed calculations of the resultant count statistics have been given by Bertolotti in Cummins and Pike (1974). Since the observations are in the form of counts at discrete intervals, the integral in the autocorrelation function

$$\varphi(\tau) = \operatorname*{Lim}_{\tau \to \infty} \frac{1}{2\tau} \int_{-\tau}^{\tau} f(t) f(t+\tau) \, dt$$

is replaced by a summation

$$\varphi'(d) = \frac{1}{m} \sum_{k=1}^{m} p_k p_{k+d} \qquad (7.10)$$

where $\varphi'(d)$ is the value for lag of d intervals of a discrete autocorrelation function evaluated over a finite time md. To find the frequencies involved one needs all values of d within a given range, but these can be processed in parallel: the difficulty is still that the evaluation for each value of d requires m multiplications, all of which must be completed in a very short time. This problem was solved in the Malvern Correlator by noting that a valid result could still be obtained if one of the series to be multiplied together was clipped to a binary form, so that multiplication required only an *and* gate for each pair of terms. The whole of the clipped series p_{k+d} from $d=1$ to $d=m$ could be made available simultaneously by passing the series of clipped p_k which are to form the clipped p_{k+d} into an m-stage shift register, each stage of which controls an *and* gate; and the unclipped p_k is fed in parallel to all these gates, so producing simultaneously all values of $\varphi'(d)$ for $d=1$ to $d=m$ (Jakeman in Cummins and Pike, 1974). The process is repeated each time a new reading is fed into the shift register and the mth term discarded. (It is because all m stages are used that the divisor in (7.10) is m rather than $m-d$.) The values of $\varphi'(d)$ are accumulated in m stores in order to average out the statistical fluctuations.

The chief application of this technique seems to have been initially to the study of liquid systems by light scattering (Cummins and Pike, 1974).

7.4 'Optical' fibre communication systems

These communication systems form a very special and limited class for practical reasons. One can assume that the radiation will not, in fact, be visible light, although the systems are described as optical, but will be in the very near infra-red, around 1 μm, and that it will be generated by solid-state lasers and detected by photodiodes, probably of the avalanche type. Signal-to-noise ratio at the input to the photodiode is required to be sufficiently good that errors in the handling of binary signals are unlikely.

If the communication channel is binary, the laser is either 'off' or is driven 'on' by a large current. The large current will be relatively noise-free because it is large and, in addition, the laser may be saturated when pulsed on, so that the light output is constant. To minimise the short but statistically varying time lag between the initiation of a current pulse from zero and the commencement of lasing, it is usual to bias the diode to the threshold of lasing rather than start from zero current. It can then be assumed that all the noise at the receiving end originates from the photon statistics, the photodiode and the amplifier. (See Section 1.7 for a limit on the ultimate performance of quantum detectors.)

The avalanche diode has an advantage over the simple PIN diode in providing some amplification of the signal before it is presented to the electronic amplifier; and, although against this must be set the additional noise associated with the multiplication process, there is some advantage in having amplification in the detector, ahead of the conventional amplifiers. The quantum efficiency of a simple PIN diode may exceed 90% and a similar figure is claimed for reach-through avalanche diodes with an anti-reflection coating on the front face and a reflecting coating on the back (Webb *et al.*, 1974). The *responsivity* is defined as the ratio of primary photo-current to incident radiation power. Because photoelectric devices give a current proportional to the number of photons, which is related to the incident power through the inverse of radiation frequency, the responsivity depends on the frequency of the incident radiation. The signal-to-noise (power) ratio at the output of a photo diode with avalanche multiplication was given in Chapter 4 as

$$\frac{S}{N} = \frac{(P_0 m R_0 M)^2/2}{2qi_{ds} + (P_0 R_0 F_s + i_{db}F_d)M^2 B + i_{na}^2} \tag{4.5}$$

where P_0 is the incident radiation peak power with depth of modulation m, i_{ds} and i_{db} the surface and bulk components of dark current, F_s and F_d represent the excess noise associated with avalanche multiplication of signal and dark currents (these factors may not be equal because the generation of these two currents may occur to various extents throughout the thickness of the diode), M the overall multiplication of photo current, i_{na}^2 the equivalent input noise current of the following amplifier and B the bandwidth. A detailed analysis of optimisation of S/N has been given by Personnick (1973) and by Wiesmann (1978): one general conclusion is that the following amplifier should present a high input impedance.

Commercial avalanche photodiodes have offered an avalanche multiplication of current of 200 to more than 600 times, a noise equivalent power of 2×10^{-13} W/Hz$^{1/2}$ for detector plus following amplifier and a radiant responsivity of 15 to 20 A/W. The dark current is 2 nA of surface current and a bulk current of 60 to 150 pA after avalanche multiplication. In a practical fibre-optical communication system, reliable detection with a silicon avalanche photodiode is obtained with 300 to 400 photons per (binary) signal pulse, while a simple PIN diode plus low-noise FET amplifier requires 1,000 to 1,500 photons per pulse (Garret, 1979). The tendency is to favour the PIN diode in practice. If the number of photons per pulse has a Poisson distribution, the inherent fluctuation in the smallest number would make a contribution to the noise of

$$\frac{(\Delta i)^2}{i^2} = \frac{(\Delta n)^2}{n^2} = \frac{1}{300} \tag{7.11}$$

The signal-to-noise ratio with an avalanche photodiode could thus be limited to 25 dB, less an allowance for detector and amplifier noise; but the PIN diode with an input of 1,000 photons per pulse would start at 30 dB.

7.5 Detectors for medium- and long-wavelength infra-red

The photons of visible light have sufficient energy to eject electrons from a suitable (low work-function) cathode into a vacuum; and in the near infra-red (around 1 μm wavelength, as used for optical-fibre communications) the photons will create hole-electron pairs in silicon, as in PIN and avalanche diodes. But at longer wavelengths all that can be done is to raise an electron to the conduction band in a narrow-gap material (e.g. 0.1 eV) and use it for photoconduction in a homogeneous device or allow it to fall across a small barrier in a junction device (photovoltaic diode). Tables 7.1 and 7.2 show the energies associated with various wavelengths and temperatures. The

Table 7.1

Wavelength (μm)	Energy per photon (eV)
1	1.24
5	0.248
10	0.124
20	0.062

Table 7.2

Temperature (K)	Energy kT (eV)
300 (room temperature)	2.58×10^{-2}
77 (liquid nitrogen)	6.63×10^{-3}
4.2 (liquid helium)	3.62×10^{-4}

photoconductor requires an external power source: it is usually operated with a relatively constant bias current. There are several measures of performance. The simplest is the *responsivity*, measured in amperes (or volts) of electrical output per watt of incident radiation. It is of this form (not dimensionless) because the theoretical response is the liberation of one electron by each photon, though in practice it may be scaled down by a quantum-efficiency factor. The rate of liberation of charge being proportional to the rate of delivery of energy is equivalent to a current proportional to the incident power. A photoconductor may actually show a 'quantum gain', i.e. the quantum efficiency is apparently greater than 100%, because the number of electrons in the conduction band depends jointly on the rate of generation and the life-time: the number of active carriers depends on the ratio of carrier life-time to transit-time through the detector. If there is a quantum gain it is because an electron reaching an electrode is replaced by another and so on until a recombination occurs after time τ. However, this also results in a slowing down of response to change in incident radiation. Most modern photoconductive cells have a response time of less than a microsecond.

But responsivity gives no indication of the signal-to-noise performance. The simplest way of allowing for this is to specify the Noise Equivalent Power. Since radiation detectors are normally used with some kind of chopper or modulator of the incident radiation, the NEP is defined in terms of the r.m.s. value of a sinusoidally modulated input, or more generally of the fundamental component of the modulated input (in case a square-wave chopper is used). The NEP is then defined as the r.m.s. value of that input which will produce an output equal to the noise. This is to be taken without restriction of the bandwidth of the noise, but it is desirable to specify the spectral distribution of the radiation used as a test signal, e.g. black-body radiation at a certain temperature. Although the input to the photodetector is a radiant *power* the output is a *current* (or voltage) so the r.m.s. measure of output noise varies as the *square root* of output bandwidth; and the alternative specification of the spectral density of NEP has units of $W/Hz^{1/2}$. The amount of radiation falling on a photodetector is proportional to its area, and if dark current can be made negligible the noise is due to incoming thermal radiation; and, allowing for the input power/output current relationship, the resulting r.m.s. noise current is proportional to the square root of area. A useful measure of performance is the *normalised detectivity* due largely to Clark Jones (1953):

$$D^* = (Area)^{1/2}/NEP \quad cm(Hz)^{1/2}W^{-1} \tag{7.12}$$

The performance of photoconductive detectors is usually limited by $1/f$ noise which one can avoid by using high enough modulation frequency. Therefore D^* is specified in terms of particular conditions of spectral character of radiation, modulation frequency and output bandwidth, the figures for which are given in brackets, e.g. $D^*(500 \text{ K}, 800, 1)$ means that black-body radiation at a temperature of 500 K was used, with a chopping frequency of 800 Hz and an output bandwidth of 1 Hz. (If the output bandwidth is specified as 1 Hz, the NEP in Equation (7.12) must be replaced by noise power spectral density, measured in W/Hz.) Table 7.3 shows the typical performance of commercial photodetectors (Mullard, 1980).

When detectors are cooled to 77 K the background output in the absence of any signal is predominantly due to any room-temperature radiation which may be incident on the detector; and, since the signal can be expected to arrive from a particular direction, the detector is

Table 7.3

Wave-length (μm)	Type	Trade name	Operating temperature	Normalised detectivity D* Conditions	Value (minimum)
2	PbS	62SV	20°C	(2.0 μm, 800, 1)	6×10^{10}
6	InSb	ORP10	20°C	(6.0 μm, 800, 1)	2×10^{8}
8–14	InSb	ORP13	77 K	(500 K, 800, 1)	5×10^{9}
8–14	$Cd_xHg_{1-x}Te$	CMT	77 K	(500 K, 5000, 1)	1.5×10^{10}

usually screened by its low-temperature enclosure from all but a limited field of view. The field of view should properly be defined as a solid angle in steradians, but it is common to express it as an angular width in degrees, e.g. 60° field of view. Photoconductive devices are usually of comparatively low resistance, tens or hundreds of ohms, so Johnson noise is not a major contribution. But the bias current produces $1/f$ noise, which is to be avoided by using a sufficiently high chopping or modulation frequency. A spectrum published by Mullard for the CMT devices shows $1/f$ noise typically commencing at a little below 2 kHz, whence the suggested modulating frequency of 5 kHz.

Another mode of operation for both mercury–cadmium–telluride and lead–tin–telluride is as a photovoltaic diode, the junction usually being formed by diffusion of a suitable dopant, e.g. excess mercury, into a thin film. The device described by Cohen-Solal and Riant (1971), with a sensitive area of 2.6×10^{-4} cm^2, was comparable with the photoconductive CMT devices in normalised detectivity, but had a response time of 2 ns as against a typical time-constant of 300 ns for a CMT device of 1.4×10^{-4} cm^2. Wang and Lorenzo (1977) tested 50 µm square elements of PbSnTe in a planar array of photodiodes produced by diffusing a dopant through windows in an oxide coating and found $D^*(11$ µm, 1,000 Hz$)$ to be 2.6×10^{10} cm Hz$^{1/2}$ W^{-1}. They also stated that $1/f$ noise was predominant below 20 Hz (contrast 2 kHz for CMT photoconductive detectors). But the unilluminated resistance of a photodiode was 66 kΩ, compared with the 20 to 200 Ω of the CMT devices, which would raise the Johnson noise enough to mask the $1/f$ noise for about $1\frac{1}{2}$ decades lower in frequency. None the less, the D^* values of the two types are comparable for similar areas; so the PbSnTe photodiode shows promise of a significant superiority over the HgCdTe photoconductor in terms of $1/f$ noise, provided such performance can be obtained consistently.

For still longer wavelengths one needs to cool the detector to a lower temperature. Wilson and Epton (1978) used indium phosphide as a photoconductive detector with peak response at 216 µm when cooled to the temperature of liquid helium. They found mostly $1/f$ noise, the spectrum having a constant slope but steeper than would correspond exactly to $1/f$ in power. With chopping at 400 Hz they found NEP $\simeq 10^{-11}$ W Hz$^{-1/2}$ and at 10 kHz, NEP $\simeq 10^{-13}$ W Hz$^{-1/2}$. They hoped that the noise could be reduced by better contact technology, but general experience of $1/f$ noise makes this doubtful.

One method of obtaining additional gain between the photoconductor and the following electronic amplifier (with its inevitable contribution to noise) is replacement of the d.c. excitation or bias current by microwave excitation (Müller and Hanke, 1979). The proposed scheme is to set the photoconductor in a microwave resonant cavity which is matched to its feed when the photoconductor is unilluminated. When the resistance of the photoconductor is lowered by radiation, some microwave power is reflected from the cavity which is now mis-matched to its feed. The 'internal gain', defined as the number of electrons sent to the microwave demodulator per incident photon, is then

$$M_0 = \eta(v\tau/l)(R/R_L)^{1/2} \tag{7.13}$$

the second factor is the quantum gain of the photoconductor (see p. 137) and the third factor is the gain in the microwave system. R and R_L are the resistances of the photoconductor without and with illumination. The experimental performance using a germinium diode as the photoconductor at 290 K, and with radiation of 1.5 μm, was a noise equivalent power less than 10^{-13} W Hz$^{-1/2}$, compared with 2×10^{-13} for a typical silicon avalanche diode plus amplifier.

7.6 Heterodyne detectors

A special technique which in principal is applicable to all types of photodetector is the optical heterodyne. This might appear to involve a piece of intellectual *legerdemain*: in all the above it was assumed that the photocurrent was *linearly* proportional to the *number of photons*, but it is now said that, since the number of photons is proportional to the square of the *electric field* in the incident radiation, all photodetectors are *square law* devices. They will therefore generate a difference frequency when illuminated simultaneously by two beams of radiation of different frequencies, namely the signal and a local oscillator. Working in terms of the combination of two oscillatory electric fields implies phase coherence, and therefore it is practically essential that both signal and local oscillation should originate from lasers. (It is best of all if both originate from the same laser, as in scattering experiments.) As well as coherence of phase, the radiation from both signal and local oscillator should have the same direction of polarisation. The basis of the advantage of heterodyne detection has been set out by Jacobs (1963). If E, E_s and E_L are the resultant, signal and local-oscillator electric fields and $\overline{E^2}$ is an average over a time long compared with the period of E_s or E_L but short compared with the period of the difference frequency,

$$\overline{E^2} = \tfrac{1}{2}E_L^2 + E_L E_s \cos(\omega_L - \omega_s)t + \tfrac{1}{2}E_s^2 \tag{7.14}$$

With the usual condition that the local-oscillator power P_L is much greater than the signal power P_s, the mean current i_{dc} through the photoconductor will be proportional to $\frac{1}{2}E_L^2$ and from (7.14) the peak value of the current at the difference frequency will be $(2E_LE_s/E_L^2)i_{dc}$. But from a photon point of view, $i_{dc}=\eta(P_L/h\nu)e$. Therefore the a.c. (difference frequency) component of photocurrent has a mean square value

$$i_{ac}^2 = 2(E_s/E_L)^2 i_{dc}^2 = 2(\eta e/h\nu)^2 P_L P_s \tag{7.15}$$

But there will be shot noise from i_{dc} according to

$$i_N^2 = 2ei_{dc}\,df = 2e^2(\eta P_L/h\nu)\,df \tag{7.16}$$

From the last two equations the signal-to-noise (power) ratio is

$$S/N = i_{ac}^2/i_N^2 = \eta P_s/(h\nu\,df) = \eta(\text{signal photon rate/bandwidth}) \tag{7.17}$$

This is the maximum value of S/N and is subject to two provisos:

(1) The transit time must not be greater than the life-time of photoelectrons, as otherwise there would be generation-recombination noise in excess of the shot noise of Equation (7.16).
(2) It is assumed that the thermal noise of the photoconductor, $i_{th}^2 = (4kT/R)\,df$ is negligible compared with the shot noise of the current excited by radiation from the local oscillator. This requires that R should not be too small, or specifically

$$\frac{1}{R} \ll \frac{e^2}{2kT}\eta\,\frac{P_L}{h\nu} \tag{7.18}$$

Putting in some numerical values, $e^2/2kT = 3 \times 10^{-18}$ for $T = 300$ K, $\eta = \frac{1}{2}$ and $h\nu \simeq 2 \times 10^{-20}$ at a wavelength of 10 μm, this should not be difficult to satisfy.

In summary, provided the local oscillator power is great enough that it generates so much current through the photodetector that the shot noise in this current drowns all other sources of noise, this increase in r.m.s. noise is exactly balanced by the increase in signal amplitude due to the square-law characteristic, both being proportional to the square root of local-oscillator power. The signal-to-noise ratio is not deteriorated by the heterodyne detector but depends only on the quantum characteristics of the signal.

7.7 Pyroelectric detectors

Certain ferroelectric materials (Chapter 3) have the property that the polarisation P of a domain changes appreciably with temperature T. (This is always to be expected when working near the Curie temperature.) The pyroelectric coefficient λ is defined as coulombs per square centimetre per degree Kelvin of temperature change, $\lambda = dP/dT \, C \, cm^{-2} \, K^{-1}$. But the surface charges resulting from polarisation are eventually neutralised by surface charges of the opposite sign, either due to internal leakage or obtained from an external source. In a pyroelectric detector these charges are obtained by current flow in an external circuit between electrodes on the element and one observes this current flow, or in practice the resulting potential difference across a resistor. In the simplest case, namely that the rate of change of temperature is large compared with the reciprocals of both the thermal and electrical time-constants, the responsivity varies inversely as frequency and the detector is said to respond to the integral of the incident radiation. This is the usual condition of use. In order to obtain a useful output from a finite element (e.g. 2 mm × 1 mm) all its domains must be 'poled' in the same direction: with a basic ferroelectric substance this must be effected by applying an electric field, and the poling will be lost if the material is heated above its Curie temperature. The most favoured material for pyroelectric detectors is triglycine sulphate (TGS) which has a Curie temperature of 49°C. But it has been shown (Keve et al., 1971) that the incorporation of about 1% of laevo-analine in the TGS crystal introduces an asymmetry in the hysteresis loop, equivalent to an internal field, which prevents any switching of domains. The addition of analine also makes the rise in dielectric constant at the Curie temperature less steep, the loss there of pyroelectric activity less abrupt and the conductivity of the material an order of magnitude lower.

Since there is no domain switching and therefore no Barkhausen-type noise, the noise equivalent circuit of a pyroelectric element is as shown in Fig. 7.1 where C_e and G_e are the capacitance and conductance of the active element, G_L and C_L the load conductance and the sum of all stray capacitances, i_{th}^2 the thermal noise generator corresponding to G_e and G_L in parallel, i_N^2 the current noise of the amplifier and V_N^2 the voltage noise of the amplifier. (G_e must include any dielectric loss in the element, as well as the d.c. conductance.) Provided the amplifier noise, especially the voltage noise, is small, i_{th}^2 is

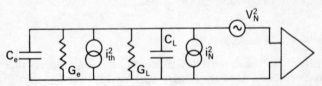

Fig. 7.1 Noise equivalent circuit for a pyroelectric detector and its amplifier

the dominant factor with G_L corresponding to a load resistor of about $10^{10}\,\Omega$. This produces white noise, but when working on the high-frequency side of the time-constant $(C_e + C_L)/(G_e + G_L)$ the r.m.s. noise voltage will vary inversely as frequency. The responsivity, volts output per watt of incident radiation, also varies inversely as frequency; but because of the watts/volts transformation the NEP is expressed in watts per square root of bandwidth. The overall effect is that the NEP of a pyroelectric detector *increases* as the *square root* of frequency between typically 1 and 1,000 Hz. At sufficiently high frequency V_N^2 dominates and the NEP then increases with the frequency. Thus a pyroelectric infrared detector should be worked with as low a chopping frequency as is practicable, e.g. 10 Hz, whereas photoconductors may have to use chopping frequencies in the kHz range to avoid $1/f$ noise. Pyroelectric detectors are heat detectors and therefore are in principle independent of wavelength. But TGS is water-soluble and so must be hermetically enclosed: the useful range of wavelengths is then fixed by the window used. A commercial detector using laevo-analine-doped TGS (Mullard RPY90) has a choice of window materials, from caesium iodide for 1 to 70 μm to sapphire for 1 to 6.5 μm and NEP(500 K, 10, 1) $= 1.0 \times 10^{-10}$ W Hz$^{-1/2}$ with wide-band windows. The recommended range of (chopping) frequency is 10 to 1,000 Hz. The great advantage of pyroelectric detectors (and of other thermal detectors) is that they operate at room temperature, even for wavelengths of 70 μm or more.

With a sufficiently strong signal to overcome noise, the frequency response of a pyroelectric element can be greatly extended. Glass (1968) recorded a rise time of 30 ns from an element of strontium barium nitrate (SBN) and Putley (1970) estimated that the response could be extended by circuit correction to about 100 MHz where the (mechanical) piezo-electric resonance was to be expected. Arrays of pyroelectric elements have been used as infra-red television cameras. In this case the surface charge is detected by scanning the array with an electron beam; and the resulting shot noise is the dominant source of noise. Assuming an amplifier noise of $2\,\text{nV/Hz}^{1/2}$ and a stray capacitance of 20 pF, a low-definition television camera (bandwidth $= = 0.5$ MHz), recording temperature differences of 4°C within its field of view would require to receive a radiation intensity of 6×10^{-5} W/cm^2 (Holeman and Wreathall, 1971).

7.8 Thermal detectors (bolometers)

The extension of the wavelength range of photoconductors to 200 μm (Wilson and Epton, 1978) which at 0.2 mm nearly meets the upper frequency limit of sub-millimetre radio techniques, and for shorter wavelengths the development of pyroelectric devices, have largely eliminated the use of purely thermal detectors; but these are reviewed for the sake of completeness and because one (the Golay cell) is singularly fundamental.

The mean-square fluctuation in the output of the detector will be the sum of the fluctuation in the radiation and the fluctuations arising within the detector, and the first source of fluctuations within a thermal detector is the fluctuation in heat content of the element which absorbs the radiation. This element should be small, so that a given amount of radiation will provide as large a temperature-change as possible, but reduction in size will increase the relative importance of thermal fluctuations. The relationship between time-constant of receiver and minimum detectable energy was investigated by Clark Jones (1947). Thermodynamics tells us that a body of thermal capacity C in contact with a large heat reservoir will exhibit temperature fluctuations of which the mean-square value is

$$\langle \delta T^2 \rangle_{\text{ave}} = kT^2/C \tag{7.20}$$

If radiation is supplying power P and the total loss of heat from the element to its surroundings is equivalent to the effect of a thermal conductance κ, the temperature-rise ΔT of the element is fixed by the relationship

$$P = \kappa \Delta T$$

and the response-time of the element may be defined as

$$\tau = C/\kappa \tag{7.21}$$

If P_m is defined as the power which makes the mean temperature-rise equal to the r.m.s. value of the fluctuation, which is a possible criterion of the least detectable radiation power,

$$P_m \tau = (CkT^2)^{1/2} \tag{7.22}$$

The product $P_m \tau$ is the effective received *energy*, and by extending the integrating time τ, one can reduce the minimum detectable power.* Obviously, one can also increase sensitivity by refrigerating the element which converts radiant energy to temperature change, so as to reduce the magnitude of $\langle \delta T^2 \rangle$ as given by (7.20). If (7.22) is re-

---

* By a rather involved argument based on the assumption that all cooling is by radiation, but none the less the rate of heat loss can be described by an equivalent thermal conductivity, Clark Jones arrived at the conclusion that sensitivity varies inversely as the *square root* only of the time-constant. In most practical cases it is reasonable to assume that heat loss is proportional to the first power of temperature-rise, leading to the proportionality to $1/\tau$ which is stated above and which has also been stated by Havens (1946).

written as

$$P_m\tau = (CTkT)^{1/2} = kT(CT/kT)^{1/2} \qquad (7.23)$$

where CT is the heat content of the element of heat capacity C when it is at temperature T, it is seen that the minimum detectable energy is the equipartition unit kT multiplied by the square root of the number of such units in the heat content of the detector, i.e. it behaves as though the heat content was made up of equipartition units and subject to $\sqrt{M}$ fluctuation. It is obvious that for any practicable size of detector body CT/kT will be very large, since it is of the order of the number of atoms in the body, and the thermal fluctuations will swamp any fluctuations in the incident radiation.

The most direct form of thermal detector is the pneumatic cell described first by Zahl and Golay (1946) and subsequently analysed in detail by Golay (1949). A cell filled with gas at reduced pressure contains an absorbing metal foil which serves to transform incoming radiant energy into an increase of gas temperature, and hence gas pressure. One wall of the cell consists of a thin membrane with a reflective coating, and this membrane is deflected by the pressure change. (Golay calculated that for the minimum detectable energy the deflection was about 10^{-10} in.) By means of a grid of alternate clear and opaque strips in front of the mirror, the intensity of reflected light is made to depend critically on the position of the membrane, and the light passing out is applied to a photomultiplier. The limit in this device is the thermal fluctuation in the cell, provided a sufficient intensity of light can be made to reach the photomultiplier. (The theoretical limits to light intensity on the mirror are (1) heating of the cell if the reflection is not perfect and (2) fluctuations in the radiation pressure on the membrane; neither of these appears to have been significant in practice.) In one cell, having an absorbing foil of area 7 mm^2, the theoretical r.m.s. fluctuation was 5×10^{-12} W in a bandwidth of 1 Hz, while the experimentally determined r.m.s. noise level was 5×10^{-11} W. The largest single contribution to the inefficiency was the leakage of heat direct to the walls of the cell, without heating the gas, and this was estimated to increase the minimum detectable power by a factor of $\sqrt{3}$; and various other sources of loss accounted for most of the rest of the factor of 10.

The other usual forms of thermal detector are the thermocouple and the resistance thermometer which is commonly known as a bolometer.* In both of these there is conversion of heat to electrical energy, so that one has to consider three types of noise: photon

---

* 'Bolometer' properly means a device for measuring radiation, regardless of the particular mechanism employed.

fluctuations, thermal fluctuations in the receiver, and the noise figure of the electrical amplifier. Havens (1946) derived the electrical power from the thermal power by writing

$$E^2/R = \alpha P\, \Delta T/T \qquad (7.24)$$

where α is a constant of the thermal-electrical conversion. The proportionality to $\Delta T/T$ must occur in a thermocouple, because a thermo-electric generator behaves like a heat engine operating in a cycle, and the theoretical efficiency is therefore the ratio of temperature difference to temperature at which heat is rejected. The relationship is less obviously fundamental for a resistance bolometer, but it is still qualitatively plausible if the resistance is a pure metal obeying the law $R \propto T$. In that case the proportion of the bridge exciting energy which is deflected to the detector is proportional to $\Delta T/T$, and the maximum bridge exciting energy is a critical factor which will be included in the coefficient α. Havens states that α is unity for thermocouples and metal bolometers, but may be as high as 100 for high-resistance bolometers.

Fellgett (1949) went further by developing an exact analogue of the Nyquist formula (a proceeding which is justified by the work of Callen and Welton, see Chapter 1) which reads

$$\langle \delta T^2 \rangle_{\text{ave}} = kT^2/C = 4kT^2 R_{\text{th}} B \qquad (7.25)$$

where C is the heat capacity, R_{th} the thermal resistance and B the bandwidth. In contrast to the electrical formula in which C would be capacitance, R electrical resistance and the left-hand side would be a voltage fluctuation, one notices temperature *squared* on the right-hand side. This is because the electrical quantity $\delta V^2/R$ is a power, but the thermal quantity $\delta T^2/R_{\text{th}}$ has the dimensions of power multiplied by temperature: $\delta T/R_{\text{th}}$ is joules per second or power. All the expressions must therefore be multiplied by an extra T when they refer to fluctuations of temperature instead of voltage. For thermocouples, Fellgett further introduced a 'dynamic resistance' which plays the same role as the radiation resistance of an aerial in the interchange of power between radiation and electric circuit, and which can be regarded as the seat of Johnson noise corresponding to the fluctuations actually arising in the electrical output of a thermocouple:

$$R_D = T\Sigma^2 R_{\text{th}} \qquad (7.26)$$

where Σ is the thermoelectric power, and then

$$\delta V^2 = 4kTR_D B$$
$$= 4kT^2\Sigma^2 R_{\text{th}} B \qquad (7.27)$$

which is obviously equivalent to (7.25). Just as is the case with the Johnson noise in the radiation resistance of an aerial, the noise seen in the dynamic resistance at the output of a thermocouple is the electrical translation of the noise in the radiation impinging on it, so that if the dynamic resistance of the thermocouple can be computed, the noise output is obtained very simply as $V^2 = 4kTR_DB$. (The Nyquist formula can be used since there is no additional power input to the thermocouple.)

7.9 Thermocouples

The thermocouple suffers the further disadvantage that the temperature rise produced by a given incident power is limited by the thermal conductance between hot and cold junctions. In metals this thermal conductance is in approximately constant ratio to the electrical conductance, according to the law of Wiedemann and Franz. Therefore all modern thermoelectric devices use semiconductors which (1) have a high value of thermoelectric power and (2) do not conform to the Wiedemann and Franz law. Schwartz (1952) used p and n junctions in series in the arrangement shown schematically in section in Fig. 7.2. The semiconductor elements were of conical shape, with the points in contact with a thin foil which could be coated on the outside to ensure maximum absorption of the radiation. This constituted the hot junction. The bases of the conical elements were connected through metal rods to a massive base which constituted the cold junction: this was divided by an insulator so that the thermo-e.m.f. could be measured. A vacuum thermocouple 2.0 mm × 0.2 mm had a time-constant of 0.03 s, a responsivity of 90 μV/μW and a resistance of 200 Ω. The minimum detectable power, set by Johnson noise, was less than 10^{-10} W for a chopping frequency of 13 Hz and bandwidth of 1 Hz.

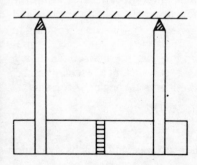

Fig. 7.2 Schwartz thermocouple

Clark Jones (1953) made various comparisons between different types of detector; and from his Fig. 3 and Section IV, one can extract the following figures for the least energy in watts detectable by various devices *if all are reduced to an area of 1 mm² and a time-constant of 0.01 s* with ambient temperature of 300°K.

Ideal heat detector	2.8×10^{-11}
Golay detector	$\simeq 10^{-10}$
Best thermocouples	3×10^{-10}
Best bolometer	6×10^{-10}

If one takes the bandwidth as $1/2\pi\tau = 16\,\text{Hz}$ for $\tau = 0.01\,\text{s}$, the ideal heat detector would have a least detectable power of $7 \times 10^{-12}\,\text{W/Hz}^{1/2}$ and the Golay cell about $2.5 \times 10^{-11}\,\text{W/Hz}^{1/2}$. This is comparable with photoconductors and photodiodes with cooling where appropriate, and the thermal devices are independent of wavelength; but they cannot achieve a fast response. They can be used for the detection of thermal sources, but their response is too slow for either signalling or scanning systems.

7.10 Cryogenic InSb bolometer

The cryogenic InSb hot-electron bolometer was mentioned in Chapter 4 and the possible sources of noise associated with it have been analysed by Brown (1984). Of the four possible sources of noise, shot noise (otherwise known as *injection noise* in semiconductors) is negligible if the contacts are of low resistance: if there is no barrier at the contact the passage of an electron through it does not produce a voltage pulse in the circuit. The conduction electrons are not 'frozen out' at 4 K because the donor levels overlap the conduction band; and correspondingly there is no g-r noise because the electrons remain in circulation. There is phonon noise representing fluctuations in power flow between electron gas and lattice caused by phonon population fluctuations. (The 'phonon population' can be regarded as a representation of the state of thermal agitation of the lattice atoms.) The spectral intensity of this phonon noise is

$$S_{\text{ph}}(f) = 4kT_1^2 G_{\text{th}} \tag{7.28}$$

where T_1 is lattice temperature and G_{th} is thermal conductance between electron gas and lattice, a conductance which can be determined through external measurements through the formula

$$G_{\text{th}} = I^2 \frac{dR}{dT_1} \frac{Z+R}{Z-R} \tag{7.29}$$

where I is bias current, R is sample resistance and Z is its dynamic resistance at zero frequency. ($Z \neq R$ because of the lag introduced by the heating effect.) Then there is Johnson noise (thermal or diffusion noise) due to fluctuations of velocity within the electron gas. One has no right to apply the Nyquist formula, because the presence of bias current indicates lack of equilibrium; and the non-equilibrium s.c.l. diode is no parallel, because there is no limitation by space charge in the bolometer. Experimentally the part of the total noise which is attributable to Johnson noise is only half that which would be predicted if one used the Nyquist formula with the value of resistance as measured in the circuit and the lattice temperature. This reduction is ascribed to electro-thermal feedback whereby the bias current does work which opposes the thermal fluctuation.

The total noise observed by Brown was of the order of 10^{-18} V^2/Hz and the resistance was of the order of 3–1 kΩ for bias currents of 8–32 μA. Philips and Woody (1982) reported that an InSb bolometer used as mixer with the low value of 10^{-6} W of local oscillator power at 500 GHz (0.6-mm wavelength) had a noise temperature of 350 K. The limitation is that the bandwidth is limited to 1 MHz by the relaxation time of the electron temperature.

7.11 Theory of gravitational radiation

The greatest challenge to low-noise techniques is the detection of gravitational waves (if they exist). The general theory of relativity (GR) predicts the existence of gravitational radiation in some circumstances, though some other post-Newtonian theories of mechanics do not. The attempt to detect gravitational waves, the success of which would provide a decisive discrimination between several theories of post-Newtonian mechanics, is of interest because of its requirement for low-noise designs owing to the extremely low level of signal expected. Although gravitational waves are predicted by GR to propagate according to a law similar to Maxwell's equations for electromagnetic radiation, they are predicted to be of quadripole form rather than dipole. In the simplest case this means that, whereas in a dipole electrical radiator charge moves from end to end, in a quadripole gravitational radiator mass moves between outside and centre, thus conserving total momentum throughout. Correspondingly the gravitational radiation is to be detected through change in the gravitational attraction between two masses, not through force on one mass. This mutual force is very small. It has been pointed out that the ratio of gravitational to electrostatic force between two electrons is $Gm^2/e^2 = 10^{-43}$, with both subject to the same inverse square law of distance. This simple comparison is not fair in practice, since we are concerned with the gravitational fields of massive objects containing also protons (and neutrons) with approximate balance of negative and positive elementary charges, so that the overall mass to charge ratio is much greater than the m/e of the electron. None the less, one can expect the effect of a typical gravitational field to be much less than that of a typical electromagnetic field. Hence the present hope is to detect the gravitational radiation from unusually energetic events, particularly from the creation of a super-nova star.

A less obvious source of gravitational radiation is a rotating body, the radiation from which is a maximum at an angular frequency equal to twice the angular velocity of rotation (Weinberg, 1972):

$$P(2\,\Omega) = \frac{32G\Omega^6 I^2 e^2}{5c^5} \tag{7.30}$$

where I is the moment of inertia and e is an eccentricity parameter. There is no radiation from a body which is rotating about an axis of symmetry, but at the other extreme $e = 1$ for an orbiting body. The factor G/c^5 is of order 10^{-53} in SI units so a large value of Ω and/or of Ie is necessary to generate appreciable radiation. It is believed that indirect evidence for such radiation has been obtained from radio-astronomy observations of the pulsar PSR 1913 + 16 (Davies, 1979; Taylor et al., 1979). This is a binary star and the slowing down of its orbital rotation at a rate of $(3.2 \pm 0.6) \times 10^{-12}$ s/s of orbital period is consistent with the expected loss of energy through gravitational radiation. (The basic theory was given by Peters, 1964.) The resulting gravitational radiation received at the earth would be too small to detect and at a much lower frequency than that to which most experimental detectors are tuned. There is therefore no prospect of direct experimental confirmation of this indirect evidence.

The form of signal to be expected is obviously crucial to the design of the detector and this question was examined by Gibbons and Hawking (1971). Weber used a massive cylinder resonant in longitudinal vibration at 1660 Hz and it is generally assumed that the signal is likely to be a disturbance lasting about a millisecond. (Gibbons and Hawking pointed out that light would travel less than 100 km in a half cycle of Weber's resonator and it is inconceivable that any remote system as small as this would create, according to formula (7.30), a detectable signal at the earth. Therefore continuous radiation from a rotating body is ruled out of consideration.) From the theory of the transformation of mass into the energy of gravitational waves Gibbons and Hawking deduced that a finite burst of energy resulting from a gravitational event must have a Riemann tensor of which the integral over the period of the burst is zero, i.e. the signal is a 'wave packet' rather than a unidirectional pulse. If the gravitational event is the capture of one mass by another, only the first integral need be zero so that a doublet pulse, approximated by one cycle of sine wave, is a possibility. If the event is the gravitational collapse of a star, both second and third integrals must be zero; and to satisfy this condition the Riemann tensor must change sign at least three times, e.g. in two cycles of oscillation. Most experimenters have employed a Q of thousands or more for the resonator, so the duration of the burst is much less than the damping time of the resonator and the change in energy as a result of the signal is likely to be nearly independent of Q. (See Section 7.12.) The mean thermal *energy* of the resonator is equal to kT but the rate of change of either *amplitude* or *phase* is proportional to Q^{-1}. One should therefore look for a coherent change in either phase or amplitude over a large number of cycles as a result of a signal, in contrast to the random changes due to Brownian motion. By observing over n cycles it should be possible to detect added energy of $2\pi n k T Q^{-1}$ and high Q is thus justified for a signal consisting of a short burst. The best which can be achieved with appropriate processing of the observational data is to make the noise contribution from the antenna equal to that from the electronic amplifier used to read out movement of the antenna, so that the effective noise temperature of the whole system is twice the noise temperature of the amplifier (Giffard, 1976).

7.12 The role of Q

The consensus of design proposals for a gravitational-wave antenna is a high-Q cylinder with a resonant frequency around 1 kHz. It is suggested that the signal might last about 1 ms and that one or more integrals of the signal must be zero, which together imply that a single cycle of sine wave could be an approximation to the signal from a 'capture' event, or two cycles for a collapse. Since large values of Q are used (5,000 to 500,000) it is highly improbable that the unknown signal will fall within the narrow pass-band of the resonant antenna, so the immediate response will be independent of Q. If a sine wave of (angular) frequency ω and period τ is applied to a resonant system of natural frequency n and damping factor n/Q the response consists of two terms:

$$\text{Forced oscillation} = \frac{F}{m\omega^2} \left\{ \frac{2\delta n}{n} \frac{\sin \omega t}{(2\delta n/n)^2 + (n^2/\omega^2)(1/Q^2)} \right.$$
$$\left. - \frac{n}{\omega} \frac{1}{Q} \frac{\cos \omega t}{(2\delta n/n)^2 + (n^2/\omega^2)(1/Q^2)} \right\} \quad (7.31)$$

$$\text{Free oscillation} = \frac{F}{m\omega^2} \frac{\omega}{n} e^{-(n/2Q)t} \left\{ \frac{-(2\delta n/n)\sin nt}{(2\delta n/n)^2 + (n^2/\omega^2)(1/Q^2)} \right.$$
$$\left. + \frac{(1/Q)\cos nt}{(-2\delta n/n)^2 + (n^2/\omega^2)(1/Q^2)} \right\} \tag{7.32}$$

We may assume that $\omega/n \simeq 1$ but Q is so high that $2\delta n/n \gg 1/Q$ with $\delta n = |n - \omega|$, so that only the sine terms remain and are independent of Q except for the damping factor in the free oscillation. The ending of the signal is represented analytically by subtracting from the signal $\sin \omega t$ another signal, $\sin \omega(t + \tau)$ (or $\sin \omega(t + 2\tau)$ for a two-cycle signal). The forced oscillation terms then cancel, leaving the residual free oscillation as the sum

$$R = \frac{F}{m\omega^2} \frac{\sin nt}{2\delta n/n} e^{-(n/2Q)t} + \frac{F}{m\omega^2} e^{-(n/2Q)(t+\tau)} \frac{\sin n(t+\tau)}{2\delta n/n}$$

Since ω is near the period of oscillation, $\sin \omega(t + \tau) \simeq \sin \omega t$ and the sum reduces to

$$R = \frac{F}{m\omega^2} \frac{\sin nt}{2\delta n/n} e^{-(n/2Q)t} \left[1 - e^{-(n/2Q)\tau}\right] \tag{7.33}$$

The merit of high Q is then the slow decay, according to $\exp(-nt/2Q)$, of the residue of the disturbance created by the signal. This makes it possible to sample the output of the detector many times (after the signal has passed) to see whether there is a component oscillation of constant phase and slowly declining amplitude super-imposed on the random output due to the thermal agitation of the cylinder. If $2Q$ is as large as 10^5 to 10^6 and the resonant frequency of the cylinder of order 10^3 Hz, the free oscillation will be appreciable for something like 100 to 1,000 cycles. If samples are taken at a rate of two per cycle (in accordance with Shannon's sampling theorem) the interpretation of a thousand or so samples is a major part of the operation of the detector.

7.13 Practical design

Assuming now that we have a detector in the form of a massive cylinder with high-Q resonance in the mode of longitudinal vibration, there are three methods of obtaining an electrical signal correspond-ing to the mechanical vibration.

(1) Weber's method was to apply strain gauges round the centre of the cylinder: during longitudinal resonant vibration the displace-ment of the ends is accompanied by strain at the centre.

(2) A logical extension of this is that the cylinder may be actually divided at the centre and the force between the two halves measured with strain gauges.

(3) The motion of the free end(s) may be detected by some electrical means such as capacitance variation. Some version of this is now usually adopted.

It is now usual to cool the cylinder to 4.2 K in order to reduce Brownian motion; and this is a substantial engineering task for a detector like the 3-m-long 5,000-kg gravitational wave detector designed by the Roma Group at CERN. A further advantage of cooling is that the mechanical damping in a metal is proportional to temperature: at liquid-helium temperature an 11.2-kg model for the Roma/CERN antenna had a Q of 5×10^5 (Rapagnani, 1982).

An important factor in the matching of antenna noise and amplifier noise is β, the proportion of the elastic energy of the antenna which can be transferred to the electrical circuit in one cycle: this is a measure of the tightness of electromechanical coupling. The input to the electronic amplifier following the transducer will, in general, behave as a source of current noise I_n^2 and voltage noise V_n^2 and the ratio of V_n/I_n to the output electrical impedance Z_0 of the transducer

$$\lambda_0 = \frac{V_n}{I_n} \frac{1}{|Z_0|}$$

should satisfy the condition $1 < \lambda_0 < \beta Q T_n/2T$ where T_n is the noise temperature of the amplifier and T the temperature of the antenna. One hopes that T_n will be much smaller than T so βQ must be large. This is a point against the split-bar antenna, since strain gauges do not give a high value of β and the split bar tends to have a much lower Q than a whole bar.

The small value of expected signal arises partly from the remoteness of its origin but even more from the low responsivity to quadripole radiation of the gravitational wave antenna, an effect which is described as a low value of the cross-section of the antenna for a gravitational wave. For the signal, Tyson and Gifford (1978) calculated that if a mass equal to 10% of the sun's mass were converted to gravitational radiation at the appearance of a supernova at the centre of our galaxy, the total energy flux (of all frequencies) reaching the earth would be about 1.5×10^4 J/m^2. Since the appearance of a supernova is not a frequent event, a more realistic approach is to look for the appearance of a supernova in one of the Virgo cluster of galaxies; and their distance is such that the expected spectral energy density at the earth would be about 10^{-4} J/m^2 Hz. This in itself is not such a low level; for if this energy is delivered in 1 ms the power level is 0.1 W/m² in a bandwidth of 1 Hz. The difficulty arises because the effective cross-section of a gravitational-wave antenna is so small and the least energy density which could be detected is kT/Σ where Σ is the cross-section of the antenna for gravitational waves impinging perpendicular to the antenna axis and T is the equivalent noise temperature of the whole detector system. The change in amplitude of vibration of a cylinder of the order of metres in length is expected in practice to be in the range 10^{-17} to 10^{-19} cm—9 to 11 orders of magnitude less than the atomic spacing in a solid or 13 to 15 orders less than a wavelength of visible light. Since the effective noise temperature of the whole system cannot be less than twice the noise temperature of the electronic amplifier, detection of this level of signal would require an amplifier T_n of about 10^{-7} K, corresponding to a noise figure which differs imperceptibly from 0 dB at an ambient temperature of a few kelvins.

7.14 Experimental detectors of gravitational radiation

The first detector was constructed by Weber, using an aluminium cylinder weighing $1\frac{1}{2}$ tons with strain gauges around its mid-plane. (See Aplin, 1972, for details.) But it is now generally considered that the sensitivity would have to be 1,000 times greater than that of Weber's detector in order to observe gravitational waves of astronomical origin and the disturbances which he observed must have had some other origin. None the less, Weber's work started the hunt for gravitational waves.

One version of method (3) which has been suggested is to use the end of the cylinder to provide part of the capacitance of a re-entrant resonant cavity, with a separation of 1.5×10^{-5} m between the two electrodes of the capacitance. Mann and Blair (1983) used a cavity made of superconducting niobium, with $Q \approx 2 \times 10^8$, for final stabilisation of the output from a 9.6-GHz room-temperature klystron which had preliminary stabilisation via a frequency-changer and 1.4-MHz discriminator. The 9.6-GHz oscillation in the cavity had a phase noise spectral density at 5 kHz from carrier of -170 to -180 dB Hz^{-1} (relative to carrier). Movement of the end of the cylinder would be translated into frequency modulation of the 9.6 GHz.

An interesting variation is described by Rapagnani (1982). In this, the counter-electrode for the capacitance to the end of the cylinder is a disc which is mechanically attached to the cylinder at its centre (though electrically insulated) and resonant in a radial mode at the same frequency as the cylinder. Fine tuning of the resonant frequency of the disc is by adjustment of the electric field between it and the cylinder, which may be up to 5×10^6 V/m* across a gap of 50 μm. According to the well-known theory of coupled resonators, the energy is transferred periodically between the two so that one is vibrating while the other is stationary. Tests were carried out on an 11.2-kg antenna, but if the same could be applied to the 5,000-kg cryogenic gravitational wave detector of the Roma group at CERN one should attain an overall effective temperature of 60 mK using a cooled FET preamplifier or 10 mK if coupled to a r.f. SQUID. Since the noise in the r.f. SQUID can be near the quantum limit, it is difficult to see how to bridge the gap between 10^{-2} K and 10^{-7} K. Tyson and Giffard (1978) suggested that the use of a SQUID for read-out could virtually eliminate the noise from the electronic amplifier, leading to a noise temperature of 10^{-4} K. They also suggested that the use of a square law output would avoid the fundamental limit to linear amplifiers (Section 1.7). But re-working Heffner's argument for square law or heterodyne operation does not seem to promise unlimited sensitivity.

REFERENCES

Aplin, P. S. (1972). 'Gravitational radiation experiments', *Contemp. Phys.*, **13**, 283–293

Brown, E. (1984). 'Non-equilibrium noise of InSb hot electron bolometers', *J. Appl. Phys.*, **55**, 213–217

Chenette, E. R., Shimada, K. and Van der Ziel, A. (1957). 'Photomultiplier tubes as standard noise sources', *Rev. Sci. Inst.*, **28**, 835–836

Cohen-Solal, G. and Riant, Y. (1971). 'Epitaxial (HgCd)Te infrared photovoltaic detectors', *Appl. Phys. Lett.*, **19**, 436–438

Cummins, H. Z. and Pike, E. R. (1974). *Photon Correlation and Light Beating Spectroscopy*, Plenum Press; New York

Davies, P. C. W. (1979). 'Gravitational radiation at last?', *Nature*, **277**, 430

Fellgett, P. B. (1949). 'On the ultimate sensitivity and practical performance of radiation detectors', *J. Opt. Soc. Am.*, **39**, 370–376

Garrett, S. G. E. (1979). 'Devices for optical fibre communication systems', *Phys. Tech.*, **10**, 77–81

Gibbons, G. W. and Hawking, S. (1971). 'Theory of the detection of short bursts of gravitational radiation', *Phys. Rev. D*, **4**, 2191–2197

Giffard, R. P. (1976). 'Ultimate sensitivity limit of a resonant gravitational wave antenna using a linear motion detector', *Phys. Rev. D*, **14**, 2478–2486

Glass, A. M. (1968). 'Ferroelectric $Sr_{1-x}Ba_xNi_2O_6$ as a fast and sensitive detector of infrared radiation', *Appl. Phys. Lett.*, **13**, 147–149

Golay, M. J. E. (1949). 'Theoretical and practical sensitivity of the pneumatic infrared detector', *Rev. Sci. Inst.*, **20**, 816–820

Havens, R. (1946). 'Theoretical comparison of heat detectors', *J. Opt. Soc. Am.*, **36**, 355 (abstract only)

---

* The dielectric strength of air is often quoted as 30,000 V/cm, or 3×10^6 V/m, but a higher field can be applied over short distances if the electrodes are free from roughness and sharp edges.

Holeman, B. R. and Wreathall, W. M. (1971). 'Thermal imaging camera tubes with pyroelectric targets', *J. Phys. D*, **4**, 1898–1919

Jacobs, S. (1963). 'The optical heterodyne', *Electronics*, **36**(28), 29–31

Jones, R. C. (1947). 'The ultimate sensitivity of radiation detectors', *J. Opt. Soc. Am.*, **37**, 879–890

Jones, R. C. (1953). 'Performance of detectors for visible and infra red radiation', *Adv. Electron.*, **5**, 2–96

Keve, E. T., Bye, K. L., Whipps, P. W. and Amis, A. D. (1971). 'Structural inhibition of ferroelectric switching in tryglycine sulphate. I. Additives', *Ferroelectrics*, **3**, 39–48

Mandel, L. and Wolf, E. (1965). 'Coherence properties of optical fields', *Rev. Mod. Phys.*, **37**, 231–287

Mann, A. G. and Blair, D. G. (1983). 'Ultra-low phase noise superconducting-cavity stabilised microwave oscillator with application to gravitational radiation detection', *J. Phys. D*, **16**, 105–113

Mullard (1980). '8 to 14 μm single-element CMT infrared detectors', *Technical Note No. 137*

Mullard (1980). 'Development sample data on pyroelectric infrared detectors', *Technical Note No. 000*

Müller, J-E. and Hawke, C. (1979). 'Noise performance of microwave-biased photoconductive detector', *Infrared Phys.*, **19**, 533–540

Personick, S. D. (1973). 'Receiver design for digital fiber optic communication systems: I', *Bell Syst. Tech. J.*, **52**, 843–874

Peters, P. C. (1964). 'Gravitational radiation and the motion of two point masses', *Phys. Rev.*, **136B**, 1224–1232

Philips, T. G. and Woody, D. P. (1982). 'Millimeter and submillimeter receivers', *Ann. Rev. Astron. Astrophys.*, **20**, 285–331

Putley, E. H. (1970). 'The pyroelectric detector', *Semiconductors and Semimetals*, **5**, 259–285

Putley, E. H. (1973). 'Modern infrared detectors', *Phys. Tech.*, **4**, 202–222

Rapagnani, P. (1982). 'Development and test at $T = 4.2$ K of a capacitive resonant transducer for cryogenic gravitational-wave antennas', *Nuovo Cimento*, **5C**, 385–408

Schwantes, R. C., Hannam, H. J. and Van der Ziel, A. (1956). 'Flicker noise in secondary emission tubes and multiplier photo tubes', *J. Appl. Phys.*, **27**, 573–577

Schwartz, E. (1952). 'Semiconductor thermocouples', *Research*, **5**, 407–411

Taylor, T. H., Fowler, L. A. and McCulloch, P. M. (1979). 'Measurements of general relativity effects in the binary pulsar PSR 1913 + 16', *Nature*, **277**, 437–440

Tyson, J. A. and Giffard, R. P. (1978). 'Gravitational wave astronomy', *Ann. Rev. Astron. Astrophys.*, **16**, 521–554

Wang, C. C. and Lorenzo, J. S. (1977). 'High-performance, high-density, planar PbSnTe arrays', *Infrared Physics*, **17**, 83–88

Webb, P. P., McIntyre, R. J. and Conradi, J. (1974). 'Properties of avalanche photodiodes', *R.C.A. Rev.*, **35**, 234–278

Weber, J. (1966). 'Observation of the thermal fluctuations of a gravitational-wave detector', *Phys. Rev. Lett.*, **17**, 1228–1230

Weinberg, S. (1972). *Gravitation and Cosmology*, Wiley; New York, p. 272

Wiesmann, T. (1978). 'Comparison of the noise properties of receiving amplifiers for digital optical transmission systems up to 300 Mbit/s', *Frequenz*, **32**, 340–346

Wilson, W. L. Jr. and Epton, P. J. (1978). 'Characteristics of a far infrared InP photoconductor', *Infrared Physics*, **18**, 669–673

Zahl, H. A. and Golay, M. J. E. (1946). 'Pneumatic heat detector', *Rev. Sci. Inst.*, **17**, 511–515

Zworykin, V. K. and Ramberg, E. G. (1949). *Photoelectricity and its Application*, John Wiley; New York

Chapter 8

Fluctuations in Charge

8.1 Introduction

The response of most electronic devices is expressed in terms of voltage or current, but there are a few devices in which charge is the relevant quantity. These include the vidicon and related camera tubes, pyroelectric imagers, charge-coupled devices, switched-capacitor networks and ultimately capacitive digital memories. It was shown in Chapter 1 that the total noise (integrated over all frequencies) in any circuit which reduces to a capacitance at high enough frequency is

$$\overline{V_{\text{tot}}^2} = kT/C \qquad (8.1)$$

corresponding to the random energy stored in the capacitance having the equipartition value corresponding to one degree of freedom,

$$\tfrac{1}{2}\overline{CV^2} = \tfrac{1}{2}kT$$

This limit is not as yet important for an isolated capacitor, such as the smallest one is likely to find in a semiconductor memory. Current optical technology can provide line widths of 1 μm or a little less, so suppose one has capacitor plates of 2 μm × 2 μm and a dielectric thickness of 0.01 μm (100 Angstrom units) with relative permittivity of 10. In round numbers (taking the permittivity constant of free space as 9×10^{-12} F/m and kT at room temperature as 4×10^{-21} J) $C = 36 \times 10^{-17}$ F (0.36 femtofarad) and $kT/C = (1/9) \times 10^{-4}$ V^2. The corresponding r.m.s. fluctuation is just over 3 mV, but it must be remembered that kT/C represents the sum of fluctuations of all frequencies; and if the bandwidth is limited, the signal/noise ratio with a working voltage of perhaps 3 V may be better than 60 dB. Current technology uses gross cell size of nearer 100 μm square per bit (chip area divided by number of bits); the area of the storage capacitor must be appreciably less, but can be much larger than 4 square μm though the dielectric is unlikely to be thinner than suggested above.

8.2 Charge-sensitive and charge-coupled devices (CCD)

The vidicon television camera and its successors such as the image orthicon and iconoscope use a photo-emissive cathode which is divided either actually or conceptually into a mosaic of small elements (corresponding to pixels), each of which acquires charge which increases as long as it is illuminated. This charge, or rather the current which produces it, is integrated over a frame period and each element is discharged periodically by a scanning electron beam, the magnitude of charge released constituting the signal. The fundamental source of noise in all photoelectric devices is the random arrival of photons, which is equivalent to shot noise in the photoelectric current. Other practical limits are more in the nature of malfunctions, e.g. the spilling of charge from one pixel to another, so that the limit of noise is not always reached. However, cameras which employ electron beams (and vacuum) are essentially thermionic devices, not solid state.

Pyroelectric devices, which also convert radiation to electric charge, were discussed in Chapter 7.

The main interest is therefore in the charge-coupled device, which is a remarkable achievement in that a packet of electrons can be transferred between a succession of potential wells in a semiconductor with negligible loss. The simplest form of CCD, as proposed by Boyle and Smith (1970) and experimentally demonstrated by Amelio et al. (1970) is now known as the surface-channel CCD or SCCD. The semiconductor, for example p-type silicon, is covered by a thin layer of insulator, on top of which are placed 'gate' electrodes; and if the gate is made increasingly positive a depletion layer (absence of holes) is formed from the surface inwards and any electrons introduced into this space, which forms a 'potential well' for electrons, will be retained in it. A number of gates may be placed along the surface and if their potential is varied cyclically, in 2, 3 or 4 phases, the well with its charge of electrons can be caused to move along the silicon in stages, e.g. to make a shift register. Note that the electrons move to each intermediate position of the well, so that the number of electron *transfers* in a shift register is the number of register stages multiplied by the number of phases in the shift cycle, which in simple structures is equal to the number of gate electrodes. This is important when one comes to consider transfer efficiency.

The disadvantage of the SCCD is that the electrons are held near the semiconductor/insulator interface and this interface contains a population of surface states which can act as traps for electrons. This means that a transfer may be incomplete, some of the electrons being trapped in surface states, and since the trapping and release are random processes this is a source of noise as well as loss. But this effect can be small. Chick *et al.* (1983) showed that transfer noise due to surface states could be deduced from transfer loss, since both had the same origin. With a 128-bit shift register they found a charge fluctuation of approximately 2.5×10^3 r.m.s. electrons for signal packets in the range 0.2 to 2.8×10^7 electrons. Unfortunately the density of surface states increased as temperature decreased, so that transfer noise could not be reduced by lowering the temperature. A palliative is the 'fat zero' technique in which the signal charge is never allowed to fall to zero, with the intention that the surface states should remain permanently filled.

The problem of surface states was overcome by the development of the bulk, or buried channel, CCD (the BCCD) by Walden *et al.* (1972). In this structure a p-type substrate is covered with a thin layer (e.g. 2 μm) of n-type silicon to which connection is made through an $n+$ diffusion which also serves as input terminal for the signal. The *n-p* junction is reverse biased and the channel for CCD charge is formed on the *n* side of the junction. (See, for example, Beynon and Lamb, 1980.) Surface states are not involved and, moreover, the electrons are now *majority* carriers in the channel. Some residual trapping in the bulk is due to impurities and defects in the crystal, which are far less numerous than surface states. But because the channel is not immediately adjacent to the dielectric layer the capacitances to the gate electrodes are smaller and the charge-carrying capacity of the BCCD is less than that of the SCCD. A point in favour of the BCCD is that, whereas the SCCD requires a minimum of three phases in the organisation of the gates to ensure movement of charge in one direction, it is possible in the BCCD to vary the thickness of dielectric under the gates and so achieve directional discrimination with a two-phase system. Apart from trapping, some noise can arise from inefficiency of transfer. If N_s is the number of electrons in the signal packet and $1 - \varepsilon$ is the efficiency of each of $2n$ transfers (for n stages), then firstly the quantity εN_s will suffer random fluctuations according to $(\delta N)^2 = \varepsilon N_s$ at each transfer and secondly the fluctuations at different transfers will be uncorrelated so that their mean square values add and at the output the r.m.s. value is

$$\Delta N_t = \sqrt{(2\varepsilon n N_s)}$$

(There are $2n$ transfers for n stages.) Since ε may be as low as 5×10^{-6} (Trowbridge, 1982), then with $n = 400$ (suitable for an imaging array with 400-line resolution) and $N_s = 10^6$, $\Delta N_t \simeq 70$ electrons r.m.s.

If charge is fed electrically to the first stage, the input circuit may be represented by a CR combination with total fluctuation (all frequencies) of mean square voltage $\overline{V^2} = kT/C$ and corresponding fluctuation of charge

$$\overline{Q^2} = C^2 V^2 = kTC$$

Expressing ΔQ r.m.s. as a number of electrons and taking T as room temperature, C in pF,

$$\Delta N_{\text{r.m.s.}} = (1/e)\sqrt{(kTC)} \simeq 400\sqrt{C_{\text{pF}}}$$

an input capacitance of 1 pF would then have an uncertainty of charge (due to thermal noise in the charging circuit) of 400 electrons r.m.s. Thornber (1974) gave a detailed analysis of noise in charge transfer devices and stated that by using a correct arrangement of diode input this could be reduced to $\frac{1}{2}\sqrt{(kTC)}$ coulombs, or about 200 electrons if $C = 1$ pF. Since there is no gain in a CCD, electrical output generates a similar noise to that arising from electrical input. However, this output noise may be reduced by the technique of *correlated double sampling*. If the voltage on the output capacitor (usually the gate capacitance of a FET transistor) is sampled as the transfer of charge commences and again as soon as the transfer is complete, the difference between the two samples is proportional to the charge and the low-frequency part of the kTC noise is eliminated. The use of this technique usually results in halving the maximum frequency of operation, so it is only employed when minimisation of noise is important.

8.3 Optical imaging

The most important use for CCDs now is for optical imaging, in which case the initial charge is subject to random fluctuations corresponding to the random arrival of photons, which may be taken to be the same as shot noise with a Poisson distribution so that the r.m.s. fluctuation ΔN of the mean number of electrons in a charge packet is $\sqrt{N}$. In an imaging system one depends on the liberation of electrons in silicon by light quanta, with a typical efficiency of 50%. Each pixel site is coupled to a stage of a CCD shift register, so that clocking of the register results in the light-generated charges from the pixels being successively transferred to the output: in one arrangement the BCCD registers are interleaved between columns of pixels and the outputs of these registers are connected to the stages of another shift register so that an element from each column in turn may be transferred to the final output, thus completing the two-dimensional scan. Such an arrangement has been described by Trowbridge (1982). Apart from photon noise, the limit in an imaging device is set by dark current, which is due to the thermal liberation of electrons in the absence of light. As a general rule, most of the components of dark noise are halved for every 5°C drop in temperature, though there are dark noise spikes which occur at specific points in the semiconductor and which decrease ohly at the rate of halving for 8°C. Trowbridge described a device 488 by 380 elements which at a temperature of $-40°C$ had a signal/noise ratio of about 2 when working at a level of about 60 electrons per pixel, and referred to a device of 244 by 190 elements which had been used in astronomy down to a noise level of 20 electrons per pixel.

To attain the ultimate in low light level (L^3) imaging one lengthens the period of integration of charge on a pixel by time delay integration (TDI). In this system each pixel of a single line is reduplicated so that in effect one has a single line scan which is repeated; and the image is moved across this array at the same speed as charge is moved from one strip to another by clocking shift registers. This means that corresponding points in successive lines of the array all 'belong to' the same point in the image and the light from that point in the image is integrated, i.e. the responses of corresponding points in all the lines are added together to form the output for that pixel (Barbe, 1976). A particular example described by Ferrier and Dyck (1980) had provision for integration optionally over 1, 4, 8, 16, 32, 64 or 128 sites and an r.m.s. noise equivalent of 80 electrons with saturation by 10^6 electrons, or 20 electrons with saturation by 2×10^5.

The development of CCD arrays for image recording has two objectives. On the one hand, the elimination of high voltage and of vacuum with its associated glass envelope is attractive for mobile television applications, as in electronic news gathering (ENG). On the other hand, there is emphasis on low light level (L^3) operation which has applications both in astronomy and in military activities. In ENG the noise is important because one wants an excellent S/N ratio in good light and reasonable performance in poor light; but in L^3 applications the interest is usually in being able to *detect* an image, at minimum S/N ratio. So noise is important to both.

8.4 Switched-capacitor networks (SCN)

The use of an operational amplifier plus capacitance to simulate an inductance led to extensive development of active filters which avoided the use of bulky inductors for low frequencies but required capacitors and resistors of CR value commensurate with the working frequency. The use of switched capacitor networks allows the use of small capacitors to simulate large resistors in a way reminiscent of Maxwell's use of a switched capacitor in one arm of a Wheatstone bridge, balanced against a resistor in the opposite arm, to find the ratio between electrostatic and electromagnetic units of charge. The simulated resistance is $R_{sim} = 1/Cf_s$ where f_s is the switching frequency. The whole network comprising capacitors, switches (FET) and operational amplifiers may be constructed on a single chip of semiconductor and there is an incentive to minimise component sizes in order to minimise chip area. Reduction of capacitance value leads to a high impedance level—Furrer (1983) remarked that 1 pF switched at 100 kHz simulates 10 MΩ—and reduction of size of the transistors in the operational amplifiers increases noise.

The possible sources of noise in a SCN are (1) the switch, (2) the operational amplifier producing white noise and (3) the operational amplifier producing $1/f$ noise. The switch is a FET; and it is assumed that this may be treated as a resistance which is a source of thermal noise with no $1/f$ noise in the conditions in which it is used, namely either fully conducting or cut off. If the switch is inserted between a signal source of internal resistance R and a capacitor C the result is a CR combination with total (all frequency) thermal noise kT/C, independent of R, so that the fact that R has either of two values at different times is at first sight unimportant. But the *spectral distribution* depends on R. It is therefore sometimes assumed (Gobet and Knob, 1983) that when the switch is open the resistance is so high, and therefore the time constant so large, that the noise is concentrated at low frequency and if it is below the pass-band of the signal it may be regarded as fluctuation in a d.c. off-set. When the switch is closed the capacitor must be charged rapidly by the sample of the signal so that the RC time constant must be short compared with the period of the highest signal frequency, f_m, with the consequence that the noise bandwidth is much greater than the signal bandwidth.

If $C = 1$ pF and $T = 300$ K, the quantity kT/C has the mean square value 40×10^{-10} V^2, giving a r.m.s. noise voltage of 63 μV for a single sampling switch. If several switches are involved, their noise contributions will be uncorrelated (because the noise arises within each FET, independently of the switching regime) and so one adds their contributions by summing squares; but one must take note of the phasing and duty cycles of the several switches, some of which may remain closed while others are open and some of which operate in parallel with fixed capacitors. Furrer (1983) analysed the SCN for a biquad cell and found about 55 μV r.m.s. of switch noise with a bandwidth of 1 MHz (set by the cut-off frequency of the operational amplifiers) with a sampling frequency of 5 kHz but only 25 μV with $f_s = 200$ kHz. With $f_s \ll f_c$ the noise varies as $f_s^{-1/2}$ but the curve flattens as the two frequencies become comparable.

If the sampling rate is related to the signal, e.g. a small multiple of the Nyquist rate of twice the highest signal frequency, the noise is said to be 'undersampled' and some of the higher noise frequencies are reflected into the signal band by aliasing. Fischer (1982) assumed perfect impulse sampling so that blocks of noise at multiples of the signal band may be aliased without loss of amplitude; and for noise bandwidth BW_n with sampling frequency f_s the number of higher-frequency blocks of noise which are aliased into the signal band would be $N = 2(BW_n/f_s)$. If BW_n is set by operational amplifier cut-off at 1 MHz while f_s is some tens of kHz, N will range up to 100. There are, however, two minor conditions which are favourable. Firstly the sampling pulse is in practice of finite width, so if the noise is white the spectral intensity in successive aliased blocks will decline as $(\sin^2 x)/x^2$ where x is scaled to the duty cycle of the sampling pulse. (See Section 8.5.) The $(\sin^2 x)/x^2$ law has been verified experimentally by various authors (e.g. Furrer and Guggenbühl, 1981) who have connected to the input a noise generator of sufficient power to drown noise from the operational amplifier(s) and observed the under-sampled switching noise. Secondly the switch noise is not white: its total (all frequency) noise power is kT/C but within this total its squared voltage is distributed as $R^2/(R^2 + 4\pi^2 f^2 C^2)$. When R is very large (switch open) the bulk of the noise is below the lower limit of the signal band and has been regarded as equivalent to a drift or slow disturbance of the d.c. off-set which can be eliminated by the correlated sampling technique in which the effective output is taken to be the difference between two adjacent samples sufficiently close for the slow disturbance to be unchanged between them (Young and Hodges, 1979). Unfortunately one cannot reduce BW_n by lowering f_c too much because at some stage the charging (or discharging) current for a capacitor must be provided by the operational amplifier and delay here would modify the transfer function of the filter.

The relative amplitudes of the three components of noise (switch, operational amplifier white noise, operational amplifier $1/f$ noise) depend to some extent on the configuration of the network in question. (A number of the earlier papers considered only a single-stage integrator which could be used as a building block in filter construction, so the question of different configurations did not arise.) Furrer (1983) examined a two-stage network based on the conditions that it was to be a low-pass filter with pole of frequency 1 kHz and $Q = 2$, with unity d.c. amplification, total capacitance on the chip of 100 pF, gain-bandwidth product of operational amplifier = 1 MHz, input equivalent r.m.s. noise voltage = $40\,\text{nV}/\sqrt{\text{Hz}}$. (This is equivalent to $R_n = 100\,\text{k}\Omega$, whereas Gobet and Knob (1983) had assumed $R_n = 1.5\,\text{M}\Omega$. If these values of R_n seem large, recall that these operational amplifiers are very small: Furrer's second configuration would require 10 capacitors, 14 FET switches and two operational amplifiers to be built into a single chip.) Furrer calculated that with two different configurations (to achieve the same transfer function) the output noise differed to the extent of about 3 dB for switch noise and 1 dB for the (predominant) amplifier white noise. (The two configurations differed in a number of switched capacitors and in one the output was taken from the first of two operational amplifiers, in the other from the second.) However, some generalisations are possible. In most cases, with both operational amplifiers and switch FETs using MOS technology, the order of importance is (1) operational amplifier white noise, (2) switch noise and (3) operational amplifier $1/f$ noise. (But Fischer (1983) found switching noise to be dominant.) Switching frequency f_s is a relatively freely disposable parameter which influences (1) and (2) through the extent of aliasing. For large degrees of undersampling (e.g. sampling below 5 kHz when $BW_n = 1$ MHz) the r.m.s. white noise voltage decreases as $1/\sqrt{f_s}$, but for large f_s (such as 200 kHz in 1 MHz) the slope is about 10 times less (curves are given in Furrer, 1983) and there would be little advantage in raising f_s above about 20 kHz. Similarly shaped curves apply to the smaller switch noise. Furrer also gives curves for amplifier noise versus f_s under the condition that the cut-off frequency of the operational amplifier is also varied so as to maintain it at five times f_s. These show a minimum r.m.s. amplifier noise of about 30 μV for $f_s \approx$ 15 kHz (and therefore amplifier cut-off at 75 kHz) compared with about 100 μV for $f_s = 5$ kHz and $BW_n = 1$ MHz.

The trend towards reducing chip area and power dissipation both lead towards an increase of noise: on the one hand, the switching noise kT/C is increased by reducing the capacitance level and, on the other hand, the amplifier noise is increased by reducing the standing current of the operational amplifiers.

One approach to the analysis of SCNs is to seek an equivalent analogue (fixed component) circuit which can be analysed by an existing computer program such as SPICE. Provided one can find a fully equivalent analogue circuit without too much effort this is a labour-saving device in the short term. The drawback is that it may tend to conceal the physical nature of the noise sources. An example of this approach is to be found in the paper by Fischer (1982).

8.5 The $(\sin x)^2/x^2$ distribution

The amplitude distribution would be $(\sin x)/x$, but noise power is measured as a mean square amplitude. Fourier series analysis shows that a sampling wave with amplitude a and duty cycle t/T is represented by a series of harmonics of the repetition frequency, of which the mth has amplitude

$$(2a/m\pi) \sin(t/T)m\pi$$

and can be put in the form

$$(2at/T)(\sin x)/x$$

where $x = m\pi t/T$. With ideal impulse sampling $t/T \to 0$, so we have also $x \to 0$ and $(\sin x)/x \to 1$ for all m.

This is Fischer's assumption, that with ideal impulse sampling all multiples of the base-band are of equal magnitude. Pictorially this means that the $(\sin x)^2/x^2$ spectrum is so stretched out that even the first half 'hoop' does not begin to decline within the working frequency range. For any other duty cycle, the frequency range corresponding to the $(\sin x)^2/x^2$ cycles depends on the duty cycle of the sampling pulse—the zeros occur at $mt/T = n\pi$ where n is an integer.

REFERENCES

Amelio, G. F., Tompsett, M. F. and Smith, G. E. (1970). 'Experimental verification of the charge-coupled device concept', *Bell Syst. Tech. J.*, **49**, 593–600

Barbe, D. F. (1976). 'Time delay and integration sensors', in *Solid State Imaging* (Ed. P. J. Jespers, F. Van der Wiele and M. H. White), Noordhoff; Leyden, pp. 659–671

Beynon, J. D. E. and Lamb, D. R. (1980). *Charge-coupled Devices and Their Applications*, McGraw-Hill; London

Boyle, W. S. and Smith, G. E. (1970). 'Charge coupled semiconductor devices', *Bell Syst. Tech. J.*, **49**, 587–593

Carnes, J. E., Kosnocky, W. F. and Levine, P. A. (1973). 'Measurements of noise in charge-coupled devices', *RCA Rev.*, **34**, 553–565

Chick, K. D., Kriegler, R. J. and Devenyic, T. F. (1983). 'Determination of transfer noise from transfer loss measurements in SCCDs', *IEEE Trans.*, **ED-30**, 64–67

Ferrier, M. G. and Dyck, R. H. (1980). 'A large area TDI image sensor for low light level imaging', *IEEE Trans.*, **ED-27**, 1688–1693

Fischer, J. H. (1982). 'Noise sources and calculation techniques for switched capacitor filters', *IEEE J. Solid State Ccts.*, **SC-17**, 742–752

Furrer, B. (1983). 'Noise analysis of switched-capacitor biquad cells' (German), *AGEN Mitteilungen*, **36**, 25–32

Furrer, B. and Guggenbühl, W. (1981). 'Noise analysis of sampled-data circuits', *IEEE International Symposium on Circuits and Systems*, 1981, vol. 3, pp. 860–863

Gobet, C. A. and Knob, A. (1983). 'Noise analysis of switched capacitor networks', *IEEE Trans.*, **CAS-30**, 37–43

Thornber, K. K. (1974). 'Theory of noise in charge-transfer devices', *Bell Syst. Tech. J.*, **53**, 1211–1262

Trowbridge, M. (1982). 'Solid state image sensors', *Electro-optics/Laser International '82 UK Conference Proceedings*, pp. 102–111

Walden, R. H., Krambiek, R. H., Strain, R. J., McKenna, J., Schriger, N. L. and Smith, G. E. (1972). 'The buried channel charge-coupled device', *Bell Syst. Tech. J.*, **51**, 1635–1640

White, M. H., Lanke, D. R., Blaha, C. F. and Mack, I. A. (1974). 'Characterisation of surface channel CCD image arrays at low light levels', *IEEE J. Solid State Ccts.*, **SC-9**, 1–13

Young, I. A. and Hodges, D. A. (1979). 'MOS switched-capacitor analog sampled-data direct-form recursive filters', *IEEE J. Solid State Ccts.*, **SC-14**, 1020–1033

Appendix I
Shot Noise

The concept of shot noise originated with the vacuum diode (Schottky, 1918) in which it was very simple. In the absence of space charge an electron with charge q which was emitted from the cathode travelled without hindrance to the anode in time τ and this electron transit constituted a pulse of magnitude q/τ. The essential ideas were that the emissions were random in time and that both emissions and transits were uncorrelated, so that the distribution in time of the pulses followed a Poisson law. In the absence of ballistic transport, the unobstructed transit between electrodes cannot apply to a solid-state device, though it can apply to transit through a short internal region such as the intrinsic region in a p-i-n diode. There is also an implicit assumption that the life-time of a carrier in a semiconductor depends only on the generation–recombination characteristics of the semiconductor and not on the time of transit of a carrier through the device. This implies that when a carrier leaves the device through a terminal the maintenance of charge neutrality ensures that it is immediately replaced by a carrier drawn from the external circuit; and the arrival of a carrier at a device terminal does not then cause any discontinuity of current. This supposes that there is a free exchange across the contacts, i.e. that they are of low resistance. If the contacts are of high resistance the motion of the carrier, and its replacement, will be more or less impeded when it reaches a contact and the abrupt change of motion will produce some degree of shot noise.

The experimental viewpoint is that micro-phenomena are always observed through their effect on a macro-circuit. This was emphasised by Rowland (1936) in connection with vacuum photocells as well as thermionic diodes; and the result in a circuit of a collection of pulses which individually occurred independently and at random was examined in the context of thermionic emission by Campbell and Francis (1946). The important relations can be expressed in the theorems of the mean response and of the variance (mean square deviation of the response from the mean) sometimes known as the theorems of the mean and the mean square:

$$\bar{y} = a \int_0^\infty s(t)\, dt \tag{I.1a}$$

$$\overline{(y-\bar{y})^2} = \overline{y^2} - (\bar{y})^2 = a \int_0^\infty [s(t)]^2\, dt \tag{I.1b}$$

where a is the rate of occurrence of the events which occur independently and at random and $s(t)$ is the response of the macro-circuit to one event. To cover the case of a mixture of different events occurring at different rates, let the integral in (I.1b) be denoted by S_n if $s(t)$ has the value corresponding to the nth kind of event, which occurs at a rate a_n. Then the combined effect has a variance

$$\overline{y^2} - (\bar{y})^2 = a_1 S_1 + a_2 S_2 + a_3 S_3 + \cdots \tag{I.2}$$

If the event in question is the transit of an electron through a vacuum diode having a CR anode circuit and y is identified as a voltage, then

$$s(t) = (q/C)\exp(-t/CR) \tag{I.3}$$

At this point one should query whether it is legitimate to use in (I.3) a macro-circuit law to describe the effect of the arrival of a single electron in the capacitor: in what sense can it be said that q decays exponentially, without violating the indivisibility of the electron charge? One attempt to escape this dilemma might be to say that the charge is quantised but the probabilities of the electron remaining for various lengths of time in the capacitor have a negative exponential distribution. However, if the exponent is taken to be $-t/CR$ this step-wise decay leads to twice the experimental value of noise (Bell, 1960). The correct interpretation depends on the fact that the capacitor contains very many electrons which are not static but are in constant thermal exchange with the resistor (Johnson noise). The arrival of one more electron will change the *average* number in the capacitor by unity, a change in the average which will relax exponentially. This question of continuous macroscopic laws applying to group phenomena which are discontinuous on the atomic scale arose in Chapter 1. In the derivation of the impedance-field method of calculating a noise which arises at various points in a device Shockley *et al.* mentioned this problem; and it arises likewise in the 'salami' method for a non-uniform semiconductor, when one goes from a summation over finite slices, each of which has macroscopic or average properties, to an integral over a path which is essentially discontinuous on the atomic scale.

The application of shot noise theory to solid-state devices is necessarily limited, owing to the absence of uninterrupted transits between electrodes. A further problem is that the principle of charge neutrality requires that apart from the creation of hole-electron pairs a new charge carrier can enter the device only as one leaves, so the 'emission' is not random. But one can visualise a junction device, e.g. a p-i-n device, in which conduction is quasi-metallic on either side of the junction but transit through the junction (possibly with collisions between carriers and lattice) is the important phenomenon and is of brief duration. If transits are uncorrelated the resulting fluctuations of current will appear as shot noise.

The major characteristic of shot noise is that its spectrum is white over frequencies for which the transit time is negligibly small; and for higher frequencies the spectral intensity falls as the inverse square of the frequency. (For details of the transition, see Bell, 1960.) An important factor in semiconductors is generation–recombination (g–r) noise. Since this arises from switching between two states—ionisation or recombination at a donor/acceptor atom in an extrinsic semiconductor, capture or release from some form of trap, direct switching between valence and conduction bands in an intrinsic semiconductor—it must have a waveform of the type known as 'random telegraph signal', the randomness being in the length of interval between switching from one state to the other. The corresponding power spectrum is known as a Lorentzian spectrum, having the form $\tau/(1 + \omega^2\tau^2)$ where τ is a characteristic time of the random intervals. This spectrum is white for $\omega^2\tau^2 \ll 1$ and falls as $1/\omega^2$ for $\omega^2\tau^2 \gg 1$ so that it is difficult to distinguish experimentally from shot noise, except that τ is usually greater for g–r noise than for shot noise so that its spectrum cuts off at a lower frequency. In some cases g–r noise appears as a distortion of an approximate $1/f$ spectrum around the frequency corresponding to $\omega\tau = 1$ while in other cases the characteristic time is so short that g–r noise appears as white noise at all but the very highest frequencies at which the device can be used.

REFERENCES

Bell, D. A. (1960). *Electrical Noise*, Van Nostrand; London

Campbell, N. R. and Francis, V. J. (1946). 'A theory of valve and circuit noise', *J. IEEE*, **93**, Pt III, 45–52

Rowland, E. N. (1936). 'The theory of the mean square variation of a function formed by adding known functions with random phases, and applications to the theories of the shot effect and of light', *Proc. Camb. Phil. Soc.*, **32**, 580–597

Schottky, W. (1918). 'Über spontane Stromschwankungen in verschiedenen Elektrizitätsleitern' (On spontaneous current fluctuations in different conductors), *Ann. Phys.*, Germany, **57**, 541–567

Appendix II

Mathematical Notes

II.1 The Poisson distribution

In historical origin the Poisson distribution was derived as the limit of a binomial distribution when the two chances ('success' p or 'failure' $1-p$) were very unequal and it therefore came to be regarded as the distribution law applicable to 'rare events'. However, 'rare events' seems a misnomer when one is considering the large number of electrons passing through a diode when a significant current flows, and in our applications it is to be considered as the distribution law for events which occur independently at random. It depends on the possibility of dividing the running variable, say time, into small intervals dt such that there is a constant probability, $\alpha\, dt$, of one event being found in dt but negligible probability of two or more events; and α is constant so that the probability of an occurrence in one interval is independent of occurrences in other intervals. (This definition applies most simply if the events are pulses which are narrow enough not to overlap. For electrons passing through a diode, the cathode may be considered as divided into a number of uncorrelated areas, each of which is small enough to have no more than one electron in transit at any instant.) One then writes the probability of exactly n events at time $t+dt$ as

$$P(n, t+dt) = P(n-1, t)\alpha\, dt + P(n, t)(1-\alpha)\, dt \tag{II.1}$$

so that

$$P(n, t+dt) - P(n, t)/dt = \alpha[P(n-1, t) - P(n, t)] \tag{II.2}$$

(Compare Feller, 1949.) Assuming that $P(n, t)$ is differentiable in t, the left-hand side of (II.2) may be taken to the limit as $dt \to 0$ so that the equation becomes

$$(d/dt)P(n, t) = \alpha P(n-1, t) - \alpha P(n, t) \tag{II.3}$$

This differential equation in t is satisfied by

$$P(n, t) = e^{-\alpha t}(\alpha t)^n/n! \tag{II.4}$$

and αt is the average number of events in t, which may be denoted by m, leading to the usual form of the Poisson distribution

$$P(n) = e^{-m} m^n / n! \tag{II.5}$$

Note that the right-hand side of (II.5) is easily differentiable in m (and hence in t) as long as n has only integral values.

A special feature of the Poisson distribution (II.5) for integral values of n is that its variance is equal to its mean. For small values of m the distribution is asymmetric in n, but for large m and n it approximates to a gaussian distribution which is subject to the condition that the variance is equal to the mean. This is the origin of the 'law of large numbers', that in any phenomenon involving large numbers of uncorrelated events the random fluctuation in number is likely to be of the order of the square root of the average number, $\sigma = \sqrt{m}$.

The kth moment about the origin can be found by weighting the nth value of $P(m)$ with n^k and summing the series from zero to infinity; and summation of the series is possible by manipulating it as a sum of exponential series (Yule and Kendall, 1948).

$$\mu'_k = e^{-vt} \left[\frac{vt}{1!} + \frac{2^k (vt)^2}{2!} + \frac{3^k (vt)^3}{3!} + \cdots \right] \tag{II.6}$$

It is naturally found that μ'_1, the first moment *about the origin* which is the mean, has the value vt; the values of the second, third and fourth moments *about the mean* $\bar{m}$ are found to be

$$\mu_2 = \sigma^2 = \bar{m}$$

$$\mu_3 = \bar{m}$$

$$\mu_4 = 3(\bar{m})^2 + \bar{m} \tag{II.7}$$

The Poisson distribution is a distribution of discrete, integral, values, but if the mean number is large (in problems related to electric currents it is usually greater than 10^6) the stepped contour of integral values must for all practical purposes be smooth and continuous. It is, in fact, approximated for large numbers by a gaussian distribution with the special restriction that the variance is equal to the mean. The measure of skewness, which is zero for the gaussian distribution, is $\mu_3 / \mu_2^{3/2}$ which for the Poisson is $(\bar{m})^{-1/2}$ and tends to zero as $\bar{m}$ increases. Similarly the fourth moment of a gaussian is $3\sigma^2$; this is approximated by the Poisson when $\bar{m}$ is negligible compared with $3(\bar{m})^2$.

Like the gaussian, the Poisson can also be derived as the limiting case of a binomial distribution for large numbers, but in the special case that one of the probabilities in the binomial becomes very small. In this derivation (Yule and Kendall, 1948) the coefficients in the terms of the binomial are manipulated by means of Stirling's approximation for factorials,

$$\ln(n!) \simeq n \ln(n) - n \qquad (II.8)$$

which is applicable provided n is large, say, greater than 100.

II.2 Series approximation of a quantum correction

The approximation to (1.13) is obtained as follows (Bell, 1960).
First expand the term $\exp(h\nu/kT)$ as a series so that

$$\bar{P} = h\nu \left[\frac{h\nu}{kT} + \frac{1}{2!} \left(\frac{h\nu}{kT} \right)^2 + \frac{1}{3!} \left(\frac{h\nu}{kT} \right)^3 + \cdots \right]^{-1} + \frac{1}{2} h\nu \qquad (II.9)$$

On taking a factor $h\nu/kT$ out of the bracket this becomes

$$\bar{P} = kT \left[1 + \frac{1}{2!} \left(\frac{h\nu}{kT} \right) + \frac{1}{3!} \left(\frac{h\nu}{kT} \right)^2 + \cdots \right]^{-1} + \frac{1}{2} h\nu \qquad (II.10)$$

If $h\nu/kT$ is small, a binomial expansion of the bracket can be ended at the second order in $h\nu/kT$:

$$\bar{P} = kT \left[1 - \frac{1}{2!} \left(\frac{h\nu}{kT} \right) - \frac{1}{3!} \left(\frac{h\nu}{kT} \right)^2 + \left(\frac{1}{2!} \frac{h\nu}{kT} \right)^2 \right] + \frac{1}{2} h\nu$$

$$= kT \left[1 + \frac{1}{12} \left(\frac{h\nu}{kT} \right)^2 \right] \qquad (II.11)$$

REFERENCES

Bell, D. A. (1960). *Electrical Noise*, Van Nostrand; London
Feller, W. (1949). 'On the theory of stochastic processes with particular reference to applications', *Proceedings of the Berkeley Symposium on Mathematical Statistics and Probability 1945, 1946*, University of California Press; Berkeley and Los Angeles
Yule, G. U. and Kendall, M. G. (1948). *An Introduction to the Theory of Statistics* (14th rev. edn.), Griffin; London

Appendix III

Rowland's (Campbell's) Theorems

The special feature of Rowland's theorems of the mean and of the mean square is that they are concerned with the response of a physical system, not with a spectrum or other mathematical abstraction. Rowland (1936) considered the case of a series of equal pulses occurring at random intervals of time, at average rate λ pulses per second, and applied to a detecting instrument having the response $s(t)$ to a single impulse. (It is assumed that the pulses are of sufficiently short duration compared with $s(t)$ that they can be treated as pure impulses.) Then the mean response to pulses of unit amplitude is

$$\bar{y} = \lambda \int_0^\infty s(t)\,dt$$

and the mean square departure from the mean is

$$\overline{(y - \bar{y})^2} = \lambda \int_0^\infty [s(t)]^2\,dt$$

In electrical terms these are the d.c. and mean-square a.c. components of the response. The most accessible reference for a proof of these theorems is Campbell and Francis (1946) and they are sometimes known as Campbell's theorems.

Rice (1944–5) extended the theorems (1) to higher moments than the second and (2) to take account of non-uniformity of pulse amplitudes. His generalised formula is

$$\mu_n = \overline{a^n} \int [|s(t)|]^n\,dt$$

where μ_n is the nth semivariant of the response and a is a weighting factor to take account of the possibility that the phenomenon may need to be divided into several classes of elementary events, each such class having its own (uniform) amplitude of pulse. Since μ_1 and μ_2 are equal to the mean and the second moment about the mean, this is equivalent to Rowland's theorems of the mean and the mean square.

REFERENCES

Campbell, N. R. and Francis, V. J. (1946). 'A theory of valve and circuit noise', *J. IEEE*, **93**, Pt III, 45–52

Rice, S. O. (1944–5). 'Mathematical analysis of random noise', *Bell Syst. Tech. J.*, **23**, 282–332, and **24**, 46–156

Rowland, E. N. (1936). 'The theory of the mean square variation of a function formed by adding known functions with random phases, and applications to the theories of the shot effect and of light', *Proc. Camb. Phil. Soc.*, **32**, 580–597

Index